Energetic Particles in Tokamak Plasmas

Energetic Particles in Tokamak Plasmas

Sergei Sharapov

CRC Press
Taylor & Francis Group
Boca Raton London New York

CRC Press is an imprint of the
Taylor & Francis Group, an **informa** business

First edition published 2021 by
CRC Press
6000 Broken Sound Parkway NW, Suite 300, Boca Raton, FL 33487-2742

and by
CRC Press
2 Park Square, Milton Park, Abingdon, Oxon, OX14 4RN

Library of Congress Cataloging-in-Publication Data
Names: Sharapov, S. E. (Sergei E.), author.
Title: Energetic particles in tokamak plasmas / Sergei Sharapov.
Description: First edition. | Boca Raton : CRC Press, 2021. | Includes bibliographical references and index.
Identifiers: LCCN 2020050054 | ISBN 9781138545540 (hardback) | ISBN 9781351002820 (ebook)
Subjects: LCSH: Tokamaks. | Plasma (Ionized gases) | Solar energetic particles. | Magnetohydrodynamics.
Classification: LCC QC718.5.C65 S55 2021 | DDC 621.48/4—dc23
LC record available at https://lccn.loc.gov/2020050054

ISBN: 978-1-138-54554-0 (hbk)
ISBN: 978-0-367-71168-9 (pbk)
ISBN: 978-1-351-00282-0 (ebk)

Typeset in Times
by codeMantra

Contents

Preface

This material is an extended reproduction of Lectures in "Energetic Particles in Tokamak Plasmas" given by the author to MSc and PhD students at the Innsbruck University (2010–2012) and at the Plasma Physics Winter School at Australian National University, Canberra (2010). The lectures are devoted to experiments and their theoretical interpretations for present-day, high-temperature tokamak plasmas, with most of the data from the largest fusion-grade plasmas of the Joint European Torus (JET, EU) and the spherical tokamak MAST (UK).

The book was written for two main categories of readers: graduate students and beginners entering control rooms of magnetic fusion machines. Students may find useful the theoretical part of the book, while colleagues starting their way as MHD or fast ion experts in control rooms may consider the book as a practical guide to interpreting energetic particle phenomena.

The results presented in the book span over a period of 25 years of the author's work on JET (EU) and on MAST at Culham Laboratory (UK). The author is very grateful to his Culham hosts Jim Hastie and Jack Connor for the great opportunity to develop his scientific career at Culham, and to Wolfgang Kerner and Geoff Cordey for welcoming him to the Analytic Theory Group at JET. From the beginning, the author has benefited enormously from working with Ambrogio Fasoli on TAE experiments on JET. Analyses and MHD modelling of TAE data with Guido Huysmans and Duarte Borba, as well as discussions with Andre Jaun on kinetic models were very useful at that time for further development.

During all these years, collaboration with Boris Breizman, Herb Berk, and Jim Van Dam from Institute for Fusion Studies (Austin, USA) was enlightening with elegant theories our efforts to interpret exciting but often puzzling phenomena in energetic ion-driven instabilities on JET and on MAST; some results from this collaboration are presented in this book. The author thanks Mikhail Gryaznevich, Tim Hender, Simon Pinches, Matthew Hole, Matt Lilley, and the MAST Team for making studies of energetic particles so exciting at Culham. The author thanks Vasili Kiptily, Barry Alper, John Wesson, Lars-Goran Eriksson, Mervi Mantsinen, Michael Fitzgerald, and the JET Team for their support and contributions to the studies of energetic particles on JET. The author also thanks his PhD supervisor Anatolii B. Mikhailovskii (Kurchatov) for his active involvement in developing the MISHKA code at JET, and Klaus Schoepf, Victor Yavorskij, Victor Goloborod'ko (Innsbruck University), Mietek Lisak, Partrik Sandquist, and Robert Nyqvist (Chalmers) for their collaboration and valuable contributions to the theory of energetic particles in tokamaks.

The author is indebted to Stuart Morris for his assistance in preparing figures for this book. The author acknowledges the permission to use figures within this book, which was kindly granted courtesy of EUROfusion or UKAEA who retain full image/copyright of those figures.

Author

Dr. Sergei Sharapov is from the Culham Centre for Fusion Energy, UK. He graduated in experimental nuclear physics from Moscow Physical Technical Institute in 1985, and did his PhD in physics and chemistry of plasmas at Kurchatov Institute of Atomic Energy, Moscow. Subsequently, he worked on the theory of non-linear waves and energetic particle-driven Alfven waves. In 1993, he moved to work on JET and the spherical tokamaks START and MAST, located at the Culham Centre for Fusion Energy, UK. Dr. Sharapov's area of interest and expertise lies in the theory, experiment, and diagnosis of energetic particles and energetic particle-driven instabilities in magnetic fusion.

1 Magnetic Nuclear Fusion in Tokamaks

1.1 INTRODUCTION

The kind of nuclear fusion energy that powers the stars and the Sun became possible on our planet in the twentieth century, first in military applications. The aim of magnetic nuclear fusion is to make fusion energy available for peaceful industrial-scale energy production [1.1]. This energy source is nearly inexhaustible, no greenhouse gases are emitted, and the radioactive waste and other dangers of nuclear power are minimised. The possibility of achieving nuclear fusion on Earth exists owing to an exceptionally large cross-section (a measure of the ability to fuse) of a nuclear fusion reaction between the nuclei of hydrogen isotopes deuterium (D), consisting of one proton and one neutron, and tritium (T), consisting of one proton and two neutrons [1.2]:

$$D + T \rightarrow He\left(3.52\,MeV\right) + n\left(14.1\,MeV\right). \tag{1.1}$$

This fusion reaction produces highly energetic helium-4 ion, He (also called alpha particle) with electric charge +2, and neutron, n. These products of the D-T fusion have birth energies in the centre-of-mass reference frame 3.52 and 14.1 MeV, respectively.

Other fusion reactions investigated in present-day magnetic fusion, which substitute much more demanding D-T operation, but may become essential in the future in their own right, are:

$$D + D \rightarrow {}^3He\left(0.82\,MeV\right) + n\left(2.45\,MeV\right), \tag{1.2}$$

$$D + D \rightarrow T\left(1.01\,MeV\right) + p\left(3.02\,MeV\right), \tag{1.3}$$

and

$$D + {}^3He \rightarrow {}^4He\left(3.67\,MeV\right) + p\left(14.67\,MeV\right). \tag{1.4}$$

Here, each of D-D reactions (1.2), (1.3) has a probability of 50%, so the total D-D fusion rate is a sum of (1.2) and (1.3), giving the total D-D fusion rate twice that of D-D neutron rate.

The reaction (1.4) involving an extremely rare gas ${}^3He$ generates alpha particle ${}^4He$ with an energy of 3.67 MeV. Because this energy is close to the birth energy of alpha particle in (1.1) of 3.52 MeV, reaction (1.4) is used to study test alpha particles when complications associated with the use of tritium and/or the flux of D-T neutrons need to be avoided.

The cross-sections of main thermonuclear reactions (1.1)–(1.4) are shown in Figure 1.1 when the colliding D ion moves at high speed (the "projectile ion"), while the second colliding ion is stationary. Figure 1.1 shows that the fusion reaction between isotopes D and T has the largest cross-section and is therefore the "easiest" one to access in experiments. In magnetic fusion, for a plasma consisting of a mixture of D and T ions at equal temperatures, the yield of thermonuclear D-T reaction is maximised at a temperature of ~20 keV.

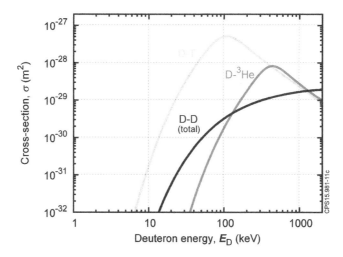

FIGURE 1.1   Cross-sections for fusion reactions D-T, D-D, and D-^{3}He as functions of the D projectile energy.

The energy released in nuclear reactions is significantly larger than that in chemical reactions because the binding energy holding a nucleus together is far greater than the energy holding an atom (electrons and ions) together. The energy gained by adding an electron to a hydrogen nucleus (proton, deuterium, or tritium) is only 13.6 eV. This value is less than one-millionth of the 17.6 MeV energy released in D-T nuclear fusion reaction. Next, in comparison with nuclear fission reactions, fusion reactions have a higher energy density. The fusion reactions produce far greater energy per unit mass even though individual fission reactions are generally much more energetic than individual fusion reactions. Only the direct conversion of mass into energy, such as that caused by the annihilation of matter and antimatter, is more energetic per unit mass than nuclear fusion.

The environmental advantages of fusion are largely determined by the type and accessibility of the reacting fuel species on Earth. The light isotope deuterium involved in all reactions (1.1)–(1.4) is naturally abundant on our planet as it constitutes 0.015% of all water. Tritium involved in (1.1) is radioactive with a half-life of 13 years, so it must be obtained first from lithium using neutron flux from a nuclear reactor via the reaction ^{6}Li+n=T+^{4}He. The raw materials for the most important D-T reaction are water and lithium, which are abundant and much less expensive than, for example, enriched uranium used in nuclear fission. The reaction (1.4), which is not a mainstream one, requires very rare ^{3}He. This can be obtained from nuclear reactors or can be found in significant quantities on the Moon.

For comparison with other fuel types, one could assess what amount of fuel is required for generating 1 GW power for 1 year (this energy is equivalent to the one typically used in a large industrial city):

Coal: 2.5 Mtonnes – produces 6 Mtonnes CO_2;
Fission: 150 tonnes U – produces several tonnes of fission waste;
Fusion: 1 tonne Li+5 ML water.

In contrast to fossil fuels, fusion does not generate "greenhouse" gases. In contrast to fission, fusion is nearly free of radioactive waste. Of course, after operating in an intense neutron flux, the structure of a fusion reactor could become activated, but the fusion fuel cycle does not generate plutonium or other long-life (thousands of years) active waste. A careful selection of materials for fusion reactor could make the activation to decay to a safe level in less than 100 years. Responsibility for the safety and integrity of a fusion facility at such a time scale could be taken without major problems as many buildings around us exist which are older than 100 years.

How can we make the nuclear forces between D and T ions work without the help of gravity existing on Sun and stars? For the fusion reaction, D and T nuclei must approach each other to a "nuclear" distance of ~10^{-13}cm. However, the nuclei are both charged positively and need to overcome the Coulomb electrostatic force between them. One of the solutions to this problem is to provide the colliding nuclei with kinetic energy larger than the Coulomb potential energy. In other words, the fuel must be hot enough with the optimum fusion rate for a DT mixture achieved at $T_D \approx T_T \approx 20$ keV (200 Mdeg). At this temperature, the DT gas is ionised and becomes a plasma – a mixture of positively charged D and T nuclei, which are the plasma ions, and negatively charged plasma electrons compensating the positive ion charge so that plasma is quasi-neutral. At this point, one needs to consider a "confinement" criterion, which is especially important for discussing plasma properties in a laboratory device. Figure 1.2 shows a sequence of the phase transitions and the relevant states of some matter, for example, an ice cube, at an increasing temperature.

All states of matter in Figure 1.2 represent different types of organisation determined by values of binding energy of that matter versus the average kinetic energy per molecule given by the temperature. If the binding energy of molecules in a crystal form exceeds the average kinetic energy per molecule, a solid state is formed. If the average kinetic energy per molecule exceeds the binding energy (a fraction of an eV), the crystal structure breaks up either into a liquid or a gas. In liquid form, when increasing kinetic energy of molecules becomes high enough to break the van der Waals forces, the liquid vaporises into a gas. Finally, when the kinetic energy of the gaseous particles exceeds the ionising potential of atoms (usually a few eV), the gas becomes plasma.

Confinement of the matter, starting from a piece of ice, becomes increasingly difficult as its temperature increases and the matter goes through the sequence of phase transitions as Figure 1.2 shows. It is easy to keep a piece of ice in hands, but a bucket is needed to confine water when this ice melts, and a balloon is required for confining water vapour when temperature further increases. It is even more difficult to confine plasma consisting of electrically charged ions and electrons. In the Sun and stars, such ionised gas is confined for millions of years by the enormous gravity of the stars, and nuclei of light elements have sufficient time to collide inside this plasma and enter into fusion reactions. However, dimensions of such fusion systems are too large to be explored on our planet. In the inertial fusion represented by thermonuclear weapons, laser, and beam fusion, the plasma is confined at a very high pressure for very short time determined by the plasma expansion. However, release of fusion energy in such short time represents a significant difficulty for integrating pulsed energy source into electric circuits delivering electricity at a nearly flat rate.

In magnetic fusion, scientists explore the key property of plasmas to conduct electricity, so that plasma can be affected by electric and magnetic fields. Confinement of plasma by external magnetic fields is the focus of magnetic fusion. It is well-known that charged particles in magnetic field move on helical orbits, that is, they circle with Larmor radius perpendicularly to the field and move freely along the field. By bending the initially straight solenoid so that the two ends of the solenoid's cylinder come together, one obtains a toroidal solenoid, as shown in Figure 1.3. Because this magnetic field toroidal topology has no open ends, charged particles flowing freely along the toroidal magnetic field move in circle Larmor orbits across the field and can remain inside the trap for a long time determined by transport processes across the magnetic field. If the toroidal loop with plasma is used as a secondary wing of a transformer, an electric inductive current I_P starts flowing in the toroidal direction as plasma is a perfect conductor. This plasma current generates a so-called "poloidal" magnetic field B_P in addition to the toroidal magnetic field B_T induced by the solenoid coils. Schematically, this concept of a toroidal magnetic field machine with toroidal plasma current

FIGURE 1.2 Schematic representation of phase transitions with increasing temperatures.

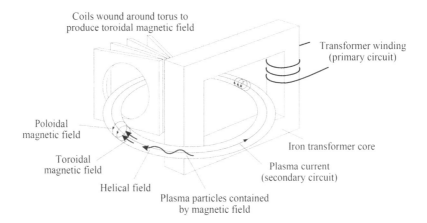

FIGURE 1.3   Schematic representation of a toroidal solenoid with plasma current used for plasma confinement in magnetic fusion.

represents a tokamak [1.3]. Tokamaks were initially conceptualised in the 1950s by Soviet physicists I.E. Tamm and A.D. Sakharov and have become popular since 1970s [1.4].

Considering D-T plasma trapped in a magnetic confinement machine and generating fusion reactions (1.1) at a reasonable rate, one would use the 14.1 MeV (80% energy) neutrons generated by D-T fusion and leaving the machine for breeding new tritium and generating the energy output, while the charged alpha particle with an energy of 3.52 MeV generated in the fusion reactions should be further confined inside the plasma. If confined well, the highly energetic alpha particles would deliver its energy to electrons and ions of the bulk plasma via Coulomb collisions. This heat flux from the fusion-born alpha particles to the plasma increases, in turn, the fusion reactivity of the plasma, and the D-T plasma becomes self-heated. When alpha particles are generated in significant numbers and the plasma self-heating by these alpha particles exceeds plasma heat losses due to radiation and thermal conductivity, the thermonuclear D-T plasma ignites. Energy production becomes possible from such "ignited" plasma. The D-T plasma could be sustained in such a state by providing relevant levels of D and T fuels and removing He ash (cooled alpha particles after transferring most of their energy to plasma).

Fusion plasmas heated by some auxiliary heating systems, in addition to the alpha particles, are called "burning" plasmas. The role of auxiliary heating is the control of plasma burn in addition to the control of D and T fuelling. This could be a more effective option for controlling non-linear exothermal plasma self-heating by fusion-born alpha particles.

1.2   SELF-SUSTAINED D-T FUSION REACTIONS AND TRIPLE-PRODUCT CRITERION FOR IGNITED PLASMA IN MAGNETIC CONFINEMENT DEVICES

There are three main conditions to obtain ignited plasma:

1. The plasma ions must be hot enough to overcome Coulomb force during the collisions between D and T ions. For D-T plasma in magnetic fusion, the temperature of plasma ions should be in the range of $T_i \approx 10\text{–}30$ keV;
2. Hot plasma must be insulated from the walls, and the plasma energy confinement time defined as $\tau_E =$ Plasma energy/Heat loss should be long enough. Here, plasma with its energy content $W = nTV$ (V is the volume of plasma) cools down as

$$\mathrm{d}W/\mathrm{d}t = -W/\tau_E \tag{1.5}$$

in the absence of any heating sources;

3. The fuel densities n_D and n_T must be high enough that fusion reactions occur at a suitable rate. Maximum plasma density is limited by impurities and instabilities.

For the particular case of D-T fusion in a magnetic camera, the criterion of plasma ignition for self-sustaining fusion reaction (similar to the Lawson criterion in inertial fusion) is expressed as a triple-product criterion [1.3]:

$$nT\tau_E > 5 \times 10^{21}\,\mathrm{m}^{-3}\,\mathrm{keV\,s}\left(\approx 10\,\mathrm{atm\,s}\right). \tag{1.6}$$

The dimension of the triple-product (pressure × time) shows that thermonuclear-grade plasma must be confined at a pressure of approximately 10 atm for 1 s. This range lies between the fusion conditions achieved in the Sun and stars, and those achieved in inertial fusion. Note here that the criterion (1.6) is only a necessary condition for ignition. It was obtained under the most optimistic assumptions.

For performing a wide range of physics experiments with fusion-grade plasmas, for example, for developing plasma heating systems and plasma scenarios, pure deuterium plasmas are often used. These generate D-D neutrons and can be used for assessing the fusion performance of DT plasma. Such D-only experiments do not require enhanced protection from 14 MeV neutrons and tritium.

Since the 1970s, significant progress has been achieved in maximising the triple-product (1.6) in equivalent D-only plasmas. This progress can be summarised as follows:

$$nT\tau_E \text{ (in equivalent D plasma)}$$

1970 – 25,000 times too small for ignition;
1983 – 100 times too small;
1995 – only five times too small.

Figure 1.4 shows how the progress in the triple-product was achieved historically, and what ion temperatures were achieved together with the triple-product values.

Two magnetic fusion machines, TFTR (United States) and JET (European Union), also conducted experiments with real D-T plasmas. The following fusion power levels in D-T plasma have been achieved:

1991 – JET – 1.7 MW (10% T; 10 MW heating)
1995 – TFTR – 10 MW (50% T; 40 MW heating)
1997 – JET – 16 MW (50% T; 22 MW heating)

We now express the ignition criterion (1.6) in engineering terms, which include magnetic field B and normalised plasma pressure $\beta = P_{\mathrm{plasma}}/P_{\mathrm{magnetic}} = 4\mu_0(nT)/B^2$ of a fusion machine. By multiplying and dividing (1.6) by B^2, we obtain

$$\beta\tau_E B^2 > 4T^2 s. \tag{1.7}$$

Expression (1.7) shows that the ignition may be achieved along three avenues:

Increasing energy confinement time τ_E;
Increasing magnetic field B;
Increasing β.

We now consider each of these options.

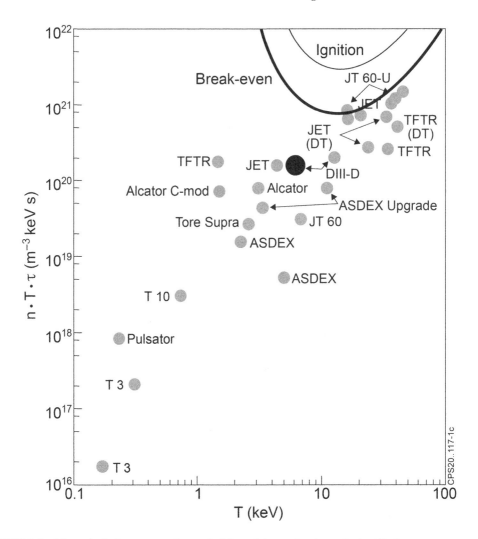

FIGURE 1.4 Magnetic fusion progress in maximising triple-product in equivalent D plasma.

1.2.1 Increasing Energy Confinement Time τ_E

Development along this avenue corresponds to a larger volume V of the fusion reactor. Indeed, when the plasma energy balance is determined by the alpha particles heating alone (in the ignited state), with the alpha particle power of $P_\alpha = 0.2\, P_{FUSION}$, for a steady-state operation of the fusion reactor, $d/dt = 0$, one obtains

$$\frac{dW}{dt} = -\frac{W}{\tau_E} + P_\alpha = 0, \tag{1.8}$$

so that

$$P_\alpha = \frac{W}{\tau_E} = nT\frac{V}{\tau_E} \tag{1.9}$$

From (1.9) we see that for a given power of P_{FUSION}, achieving ignition by increasing τ_E means a machine with a larger volume V as the plasma temperature and density are optimised for fusion conditions and neither could be varied significantly for increasing τ_E. In particular, for a magnetic fusion reactor generating $P_{FUSION} = 1$ GW (typical value of a large industrial city), the volume of the D-T plasma should be approximately $V \approx 1000\,m^3$. It follows from this estimate that:

A. The next step international project ITER [1.5,1.6] will approach the critical volume required for the ignition very closely;

B. Among all magnetic fusion machines in operation, Joint European Torus (JET) [1.7] has the largest volume of $V \approx 100\,m^3$, which is an order of magnitude below the ignition-size volume, as shown in Figure 1.5. This means that present-day magnetic fusion experiments are being developed with subcritical volumes. Such fusion research is similar to fission research if the critical mass could not be achieved.

1.2.2 INCREASING MAGNETIC FIELD B

It is technologically challenging to obtain $B > 5$ T. The engineering constraints on the coil's structural integrity generating the magnetic pressure $B^2/2\mu_0$ become very severe as the magnetic pressure is ≈ 1 kg/cm^2 for $B = 0.5$ T, but it becomes ≈ 400 kg/cm^2 for $B = 10$ T. Several projects were developed aiming at increasing B, including the present-day tokamak Alcator C-MOD (United States). Several next-generation machines were also designed, for example, IGNITOR (Italy) and FIRE (United States). Recent development of high-temperature super-conducting magnets may significantly enhance this avenue of magnetic fusion development in the future.

1.2.3 INCREASING BETA

The beta parameter shows how much plasma pressure per unit magnetic pressure could be confined in a magnetic fusion machine. The maximum value of this parameter is limited by magneto-hydrodynamic (MHD) instabilities, typically at a level of few percent, in present-day machines. In contrast to the technological difficulties in the first two avenues of approaching the triple-product ignition criterion, high volume and high B, the beta limit is determined by laws of physics. Instead of the technological improvements, an optimisation of the magnetic field topology could be the key to achieving higher values of beta.

The search for the best topology of magnetic field in tokamaks capable of achieving high beta resulted in the concept of a spherical tokamak (ST) with comparable minor and major radii, $R/a \approx 1$,

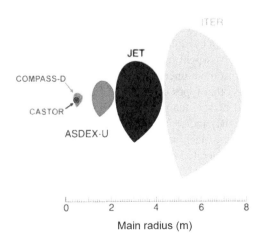

FIGURE 1.5 Comparison of plasma cross-sections for some of the machines on the way to ITER.

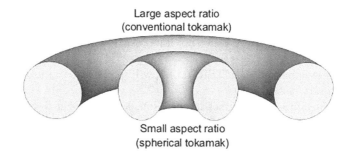

Large aspect ratio
(conventional tokamak)

Small aspect ratio
(spherical tokamak)

FIGURE 1.6   Comparison of small and large aspect ratio toroidal geometries.

as shown in Figure 1.6. Such machines achieved volume averaged beta up to $\langle\beta\rangle \approx 40\%$, which is an order of magnitude higher than the machines with typically large aspect ratio of $R/a \approx 3$. The advantages of the ST concept in achieving high beta are demonstrated in present-day machines MAST (United Kingdom) and NSTX (United States), and the next step project STEP (United Kingdom) is currently being developed.

1.3   THE ROLE OF ENERGETIC IONS IN MAGNETIC FUSION

The ultimate aim of the magnetic D-T fusion is self-heating of thermonuclear plasma by fusion-generated alpha particles, so the role of energetic alpha particles in magnetic fusion is key. A good confinement and predictable cross-field transport of these alpha particles are absolutely essential for successful fusion, so it is important to investigate in depth the transport processes involving fusion-generated alpha particles that heat the plasma. Energy of alpha particles born in D-T fusion reactions, 3.52 MeV, exceeds the temperature of thermal ions more than hundred times. Therefore, alpha particle populations that will have much lower density than the thermal plasma, $n_\alpha/n_e < 1\%$, in DT plasma experiments like ITER, will have its energy content, $n_\alpha E_\alpha$, about ~10%–15% of the thermal plasma energy content. This implies that alpha particles could notably affect MHD instabilities driven by thermal plasma pressure, and could excite new types of kinetic instabilities associated with alpha particle motion at speeds much higher than those of the thermal ions. Instabilities of weakly damped toroidal Alfvén eigenmodes (TAEs) are of particular concern as alpha particles have birth velocity exceeding Alfvén speed. During the slowing down, alpha particles enter resonance with TAEs and may excite TAEs due to the free energy source in alpha particle pressure gradient. If TAE amplitudes driven by alpha particles do not saturate at a negligibly low level, TAEs could significantly affect the cross-field transport of alpha particles; thus, making predictions of alpha particles in fusion plasmas more uncertain and difficult.

Apart from alpha particles, other types of energetic particles are often used in auxiliary heating of plasma to increase thermal plasma temperatures. Plasma heating with energetic ions produced from neutral beam injection (NBI) and/or ion cyclotron resonant heating (ICRH) are effective techniques widely used in present-day machines.

Finally, additional currents generated by energetic particles under certain conditions are often a valuable asset for controlling magnetic field topology in the plasma.

All these issues will be discussed further in this book. We end this introductory section with Table 1.1 demonstrating which parameters of fusion alpha particles and auxiliary heating systems have been achieved in present-day experiments, and how these compare with similar parameters expected on ITER (Table 1.1).

TABLE 1.1

Fast Ion Parameters for Various Tokamak Heating Systems

Parameter Machine	$P_f(0)$ (MW/m³)	$n_f(0)/n_e(0)$ (%)	$\beta_f(0)$ (%)	$\langle\beta_f\rangle$ (%)	Max $\lvert R\nabla\beta_f\rvert$	$V_f/V_A(0)$
NBI/TFTR[a]	3	13	0.9	0.4	0.04	0.35
ICRH/JET[b]	1–3	1–10	1–3	0.5	≈0.1	≈1–2
Alpha particles TFTR[a]	0.3	0.3	0.26	0.03	0.02	1.6
Alpha particles JET[c]	0.12	0.44	0.7	0.12	0.035	1.6
ITER-99[d]	0.3	0.3	0.7	0.2	0.06	1.9

[a] Parameters in TFTR DT discharge #76770 with 40 MW of 100 keV NBI, $B_T=5$ T.

[b] Typical parameters achieved in JET deuterium plasmas with up to 15 MW of ICRH ³He minority heating with ³He tail temperature $\langle E_f\rangle \approx 1$ MeV.

[c] Alpha particle parameters in JET record high fusion power (16.1 MW) H-mode DT discharge #42976 with 22 MW of NBI and 3.1 MW of ICRH, $B_T=3.6$ T.

[d] The anticipated alpha particle parameters of the "Ignited ITER" Project considered before 2000 [1.8].

REFERENCES

1. L.A. Artsimovitch, *Controlled thermonuclear reactions*, Gordon and Breach, New York (1974).
2. G.H. Miley et al., *Fusion Cross-Sections and Reactivities*, Univ. of Illinois Nucl. Eng. Report COO-2218-17, Urbana, IL (1974).
3. J. Wesson, *Tokamaks*, Oxford University Press, Oxford, 4th Edition (2011).
4. V.D. Shafranov, *Physics - Uspekhi* **44** (2001) 835.
5. ITER Physics Basis, *Nucl. Fusion* **39** (1999) 2137.
6. Progress in the ITER Physics Basis, *Nucl. Fusion* **47** (2007) S1.
7. P.H. Rebut et al., *Proceed. of the 10th Intern. Conf., Plasma Physics and Controlled Nuclear Fusion, London* (IAEA, Vienna, Vol. I, 1985), p. 11.
8. J. Jacquinot et al., *Nucl. Fusion* **39** (1999) 2471.

2 Charged Particle Orbits in Magnetised Plasma

2.1 LARMOR ORBITS IN HOMOGENEOUS MAGNETIC FIELD

Consider the motion of a charged particle of mass m and charge Ze (where $-e$ is the electron charge) in a magnetic field B, which is constant in space and time. The equation of motion for such a particle has the form

$$\frac{d\boldsymbol{v}}{dt} = \frac{Ze}{mc}(\boldsymbol{v} \times \boldsymbol{B}), \tag{2.1}$$

so the vector of particle acceleration lies in the plane perpendicular to B. The velocity of the particle parallel to B does not depend on the magnetic field and remains constant. Clearly, the parallel, $K_{\parallel}$, and perpendicular, $K_{\perp}$, components of the kinetic energy of the particle are constants,

$$K_{\parallel} = \frac{1}{2}mv_{\parallel}^2 = \text{const}, \tag{2.2}$$

and

$$K_{\perp} = K - K_{\parallel} = \frac{1}{2}mv_{\perp}^2 = \text{const}, \tag{2.3}$$

as a constant magnetic field cannot change the kinetic energy of particle, $K = \text{const}$.

For finding the velocity perpendicular to B, we introduce the Cartesian coordinate system with its axis z directed along B, so that the components of the velocity along x and y satisfy in accordance with (2.1):

$$\frac{d}{dt}v_x = \frac{ZeB}{mc}v_y, \tag{2.4}$$

$$\frac{d}{dt}v_y = -\frac{ZeB}{mc}v_x. \tag{2.5}$$

By introducing a complex value of

$$v^* \equiv v_x + iv_y, \tag{2.6}$$

one can combine two equations (2.4) and (2.5) to obtain

$$\frac{d}{dt}v^* = -i\frac{ZeB}{mc}v^*. \tag{2.7}$$

The solution of (2.7) has the form

$$v^* = \text{const} \cdot \exp(-i\omega_B t), \tag{2.8}$$

which describes a circle orbit of the particle in the plane perpendicular to $\boldsymbol{B}$ rotating with angular frequency

$$\omega_B = \frac{ZeB}{mc}. \tag{2.9}$$

This frequency is called *cyclotron frequency* or *gyro-frequency*. For electrons, the value of this frequency is

$$|\omega_{\text{Be}}| = \frac{eB}{mc} \approx 1.7 \cdot 10^7 B[G], \tag{2.10}$$

and for ions with charge Ze and mass $m = M = AM_H$, it is

$$\omega_{\text{Bi}} = \frac{ZeB}{Mc} \approx \left(\frac{Z}{A}\right) \cdot 10^4 B[G]. \tag{2.11}$$

Here, Z and A are the charge and mass numbers of the ions, respectively, M_H is the hydrogen ion (proton) mass, and magnetic field is measured in Gauss of the CGS units.

The motion of a charged particle in a homogeneous constant magnetic field is a super-position of the cyclotron rotation ω_B across the magnetic field and free motion along the magnetic field with velocity $v_\parallel$. The cyclotron frequency depends on the value of the magnetic field, while the radius ρ of the cyclotron orbit (also called *Larmor radius*) is determined by the balance between the centrifugal and the magnetic forces,

$$\frac{mv_\perp^2}{\rho} = \frac{Ze}{c} v_\perp B, \tag{2.12}$$

and depends on the value of the perpendicular velocity of the particle:

$$\rho = v_\perp / \omega_B. \tag{2.13}$$

For thermal motion of particles in the plane perpendicular to $\boldsymbol{B}$, the Larmor radii of such particles have a distribution similar to that of the thermal velocities, that is, Maxwell distribution.

By considering the momentum conservation law for cyclotron rotation, that is,

$$mv_\perp \rho = \frac{mv_\perp^2}{\omega_B} = \frac{m^2 v_\perp^2 c}{ZeB} = \text{const}, \tag{2.14}$$

the value of $v_\perp^2 / B$ is conserved, thus, implying a conservation of the *magnetic moment* of the Larmor orbit

$$\mu \equiv \frac{mv_\perp^2}{2B} = \text{const}. \tag{2.15}$$

The magnetic moment is in the direction of the unit vector normal to the area covered by the current loop due to the circular motion of a charged particle. In the present case, the direction of μ is anti-parallel to the direction of $\boldsymbol{B}$:

$$\mu = -\frac{mv_\perp^2}{2B^2} \boldsymbol{B}, \tag{2.16}$$

so that a plasma is diamagnetic.

The conservation law (2.15) is valid for particle motion in a homogeneous stationary magnetic field. In a more general case of a weakly varying magnetic field in time and space, this conservation law is approximately valid. If the variation rate and inhomogeneity of the magnetic field in time and/or space satisfy

$$\frac{\mathrm{d}\ln(B)}{\mathrm{d}t} \ll \omega_B, \tag{2.17}$$

$$\rho \frac{\mathrm{d}ln(B)}{\mathrm{d}x} \ll 1, \tag{2.18}$$

these slow variations of the magnetic field are called *adiabatic*. The magnetic moment is a nearly conserved quantity for such a system, and is called *adiabatic invariant*.

2.2 DRIFT MOTION OF LARMOR ORBITS IN THE PRESENCE OF EXTERNAL FORCES AND INHOMOGENEOUS MAGNETIC FIELD

If a not very dense plasma is affected by a strong external force F, one could neglect the interaction between the particles and consider plasma as a sum of independently charged particles moving along their own orbits determined by the externally applied forces. The only internal plasma field that could be never neglected is the electric polarisation field resulting from the charge separation and delivering the quasi-neutrality condition. For the motion of a charged particle in the presence of given external forces, one has

$$m\frac{\mathrm{d}v}{\mathrm{d}t} = \frac{Ze}{c}(v \times B) + F. \tag{2.19}$$

Equation (2.19) has a vector form and can be solved analytically in only some simple cases. In general cases, an approximate technique called *drift approach* is applied to describe the solutions of (2.19). The drift approach is valid for plasmas in the presence of external forces strong enough for neglecting the interaction between the particles of the plasma. Under these conditions, the motion of a charged particle could be decomposed into three contributions:

1. Fast cyclotron rotation in the plane perpendicular to B with gyro-frequency (2.9),
2. Drift motion of the Larmor circle across B,
3. Particle motion along B not affected by the Lorentz force.

The motion of the Larmor circle of the particle forms a so-called *guiding centre* trajectory *kinetic* or guiding centre approach to describing the particle orbits (Figure 2.1).

We represent the external force as a sum of the forces along the magnetic field and perpendicular to it:

$$F = F_\perp + F_\parallel b, \tag{2.20}$$

so that (2.19) has the following projections in the Cartesian coordinate system:

$$\frac{\mathrm{d}}{\mathrm{d}t}v_x = \frac{ZeB}{mc}v_y + \frac{F_x}{m} \equiv \omega_B v_y + F_x/m, \tag{2.21}$$

$$\frac{\mathrm{d}}{\mathrm{d}t}v_y = -\frac{ZeB}{mc}v_x + \frac{F_y}{m} \equiv -\omega_B v_x + F_y/m, \tag{2.22}$$

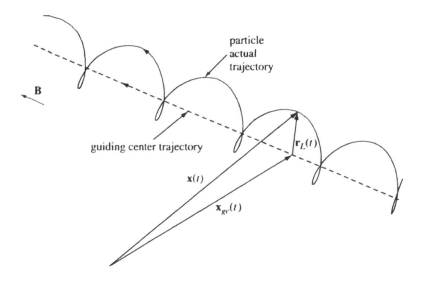

FIGURE 2.1   Relationship between guiding centre trajectory and actual trajectory.

$$\frac{\mathrm{d}}{\mathrm{d}t} v_{\parallel} = F_{\parallel}/m. \tag{2.23}$$

Here, Eq. (2.23) for the parallel particle motion describes acceleration/deceleration of the particle along B depending on the directivity of $F_{\parallel}$, and a free motion of the particle along B if $F_{\parallel} = 0$. This equation is decoupled from the equations for the particle motion across B and corresponds to the above introduced contribution (3) of the particle motion.

Equations (2.21) and (2.22) describing the motion of the charged particle across B could be represented via the complex variable $v^* \equiv v_x + iv_y$ as follows:

$$\frac{\mathrm{d}}{\mathrm{d}t} v^* = -i\omega_B v^* + \left(F_x + iF_y\right)/m. \tag{2.24}$$

The solution of (2.24) is a sum of the general solution of the homogeneous equation and the particular solution of the inhomogeneous equation,

$$v^* = \mathrm{const}\cdot\exp(-i\omega_B t) + \frac{F_y}{m\omega_B} - i\frac{F_x}{m\omega_B}, \tag{2.25}$$

provided the force $F_{\perp}$ is constant in both time and space. One can see that (2.25) consists of a fast cyclotron rotation and drift velocity which is determined by the force applied. In the vector form, the drift velocity is

$$v_{\perp D} = \frac{c}{ZeB^2}(F_{\perp} \times B). \tag{2.26}$$

2.2.1   DRIFT MOTION IN A CONSTANT ELECTRIC FIELD

Suppose now that there is a constant electric field E present in the plasma in addition to the constant magnetic field B. The relevant force in (2.19) is in this case

$$F = ZeE. \tag{2.27}$$

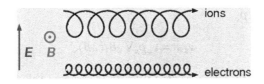

FIGURE 2.2   Larmor orbits plus the perpendicular $E \times B$ drift in a presence of an electric field.

The field component $E_{\parallel}$ along the magnetic field will accelerate the charged particle along B, while the effect of the electric field component perpendicular to the magnetic field $E_{\perp}$ will cause an electric drift velocity of Larmor circles determined by (2.26) and (2.27)

$$v_{\perp E \times B} = \frac{c}{B^2}\left(E_{\perp} \times B\right). \tag{2.28}$$

Because the electric drift velocity (2.28) does not depend on the charge or mass of the particle, in the presence of a constant electric field, both electrons and ions drift together in the same direction and with the same drift velocities. No net current is produced due to the drift in a constant electric field. Figure 2.2 shows how the super-position of the gyro-motion and drift due to electric field (2.28) looks like in the plane perpendicular to B, with the magnetic field vector directed towards us.

It may be noted that (2.28) is only valid for

$$|E| \ll |B|, \tag{2.29}$$

which is satisfied in most cases to be considered in magnetic fusion.

2.2.2   DRIFT MOTION IN INHOMOGENEOUS STATIC MAGNETIC FIELD

A precise description of charged particle orbit in inhomogeneous magnetic field usually requires a significant computational effort. Here, we consider two most important cases, both could be solved in a rather simple manner.

First, consider a static magnetic field that is no longer spatially uniform in *its magnitude*. When the magnetic field does not vary significantly over the Larmor radius, that is, when (2.18) is valid, the particle gyro-centre drifts due to ∇B. Let us split the Larmor motion in the inhomogeneous magnetic field into two parts, one of which has a smaller radius $\rho = \rho_1$ in the stronger field area, while the other one has a larger $\rho = \rho_2$ in the weaker field area. If half of the cyclotron rotation is done with ρ_1, and the other half with ρ_2, then the averaged position of the Larmor circle shifts by $\Delta x = 2\left(\rho_2 - \rho_1\right) = 2\Delta\rho$ in one cyclotron period $2\pi / \omega_B$ thus making drift velocity $|v_{\perp B}| \sim \omega_B \Delta\rho / \pi$. Taking into account that the Larmor radius varies not in a single step but gradually, the shift of the Larmor circle can be estimated more accurately as

$$\Delta x = \pi \Delta\rho_g, \tag{2.30}$$

where the variation of the Larmor radius over its length is

$$\Delta\rho_g = \rho_g \nabla_{\perp}\rho_g = -\left(\rho_g^2 / B\right)\nabla_{\perp} B. \tag{2.31}$$

The ∇B drift velocity takes the form

$$|v_{\perp B}| = \omega_B \rho_g^2 \nabla_{\perp} B / (2B), \tag{2.32}$$

or, by substituting $\omega_B = v_\perp / \rho_g$,

$$|v_{\perp B}| = v_\perp \rho_g \nabla_\perp B / (2B). \tag{2.33}$$

Equation (2.33) determines the gradient drift velocity in absolute value but not in directivity. For the component of the magnetic field gradient in the direction perpendicular to $\boldsymbol{B}$, we represent $\nabla_\perp B = \nabla B \times \boldsymbol{b} = -\boldsymbol{b} \times \nabla B$, where $\boldsymbol{b} \equiv \boldsymbol{B}/B$ is the unit vector along the magnetic field. By substituting this expression and (2.13) in (2.33), one obtains the gradient drift velocity in the vector form:

$$\boldsymbol{v}_{\perp B} = c \frac{mv_\perp^2}{2ZeB^3} (\boldsymbol{B} \times \nabla B). \tag{2.34}$$

We note that $\boldsymbol{v}_{\perp B}$ depends on the particle charge in contrast to $\boldsymbol{v}_{\perp E \times B}$ given by (2.28). In particular, $\boldsymbol{v}_{\perp B}$ is in opposite directions for ions and electrons and gives rise to a current in plasma.

Here, (2.34) could be obtained from (2.16) and (2.26) via the force

$$\boldsymbol{F} = -\mu \nabla B. \tag{2.35}$$

Figure 2.3 shows how the particle motion combining its gyro-motion and drift due to ∇B looks in the plane perpendicular to $\boldsymbol{B}$, with the magnetic field vector directed towards us.

2.2.3 Drift Motion in Static Magnetic Field with Curvature

Here, we consider the second simple case when the *direction* of a static magnetic field varies. In this case, the centre of Larmor circle moves along the curved magnetic field line with some curvature radius $\boldsymbol{R}$, so a centrifugal force arises:

$$. \boldsymbol{F} = \frac{mv_\parallel^2}{R^2} \boldsymbol{R}, \tag{2.36}$$

which is directed along the curvature radius.

By substituting (2.36) in the general expression (2.26), we obtain the expression for the curvature-B drift velocity:

$$\boldsymbol{v}_{\perp c} = c \frac{mv_\parallel^2}{ZeR^2B^2} (\boldsymbol{R} \times \boldsymbol{B}). \tag{2.37}$$

Here, $\boldsymbol{v}_{\perp c}$ depends on the particle charge, as in the case of ∇B drift, but opposite to the case of $E \times B$ drift. Hence, ions and electrons in curved magnetic field move in opposite directions and induce drift currents causing charge separation.

Note that (2.37) depends on the parallel velocity of the particles, in contrast to (2.34) depending on the perpendicular velocity. For charged particles injected along the magnetic field, $\frac{v_\parallel}{v_\perp} \gg 1$, the

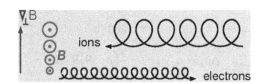

FIGURE 2.3   Particle motion consisting of Larmor orbits plus the perpendicular ∇B drift in a static inhomogeneous magnetic field.

value of ∇B drift could be small, while the value of the curvature drift (2.37) is large. This is often essential for describing charged fast ions produced from a neutral beam injection parallel to the magnetic field.

2.2.4 DRIFT MOTION IN TIME-DEPENDENT ELECTRIC FIELD

We saw in Section 2.2.1 that a constant electric field perpendicular to the magnetic field causes both electrons and ions to drift at the same drift velocity (2.28) so that no net electric current is generated in the plasma. However, a *time-dependent* electric field causes a drift perpendicular to B that depends on the mass and charge of charged particles and, consequently, produces a current in the plasma consisting of ions and electrons. We start from equation

$$m\frac{\mathrm{d}\boldsymbol{v}_\perp}{\mathrm{d}t} = Ze\left(\boldsymbol{E}(t) + \frac{1}{c}(\boldsymbol{v}_\perp \times \boldsymbol{B}) \right),$$

(2.38)

and represent the perpendicular velocity as a sum of the velocity $\boldsymbol{v}_\perp^*$ of cyclotron rotation with gyro-frequency (2.9), as well as the perpendicular drift velocity of the guiding centre. The drift velocity can be represented as

$$\boldsymbol{v}_{\perp D} = \boldsymbol{v}_{\perp E \times B} + \delta \boldsymbol{v}_\perp,$$

(2.39)

where $\boldsymbol{v}_{\perp E \times B}$ is given by Eq. (2.28), and the additional drift $\delta \boldsymbol{v}_\perp$ is to be found. If the characteristic time of the electric field variation is much longer than the inverse gyro-frequency, and $\left| \delta \boldsymbol{v}_\perp / \boldsymbol{v}_{\perp E \times B} \right| \ll 1$, we obtain

$$\delta \boldsymbol{v}_\perp = \frac{mc^2}{ZeB^2} \cdot \frac{\partial \boldsymbol{E}}{\partial t}.$$

(2.40)

The drift determined by (2.40) is called the "polarisation drift." It gives drift velocities in opposite directions for charges of opposite signs and, consequently, provides polarisation current density.

2.3 DRIFT MOTION OF ENERGETIC PARTICLES IN TOKAMAK

The axi-symmetric toroidal magnetic configuration has both radial gradient and curvature of the magnetic field, therefore, a charged particle motion is initially determined by the super-position of the magnetic drifts. However, in the absence of a poloidal magnetic field (no toroidal plasma current), the charge-dependent magnetic drifts give rise to a charge separation resulting in a vertical electric field. An outward $\boldsymbol{E} \times \boldsymbol{B}$ drift of both electrons and ions follows, expelling the whole plasma out of the high field side of the torus to the outer wall. Figure 2.4 shows the directions of the magnetic and $\boldsymbol{E} \times \boldsymbol{B}$ drifts of electrons and ions in a cross-section of a toroidal axi-symmetric configuration.

To avoid the outward electric drift of plasma, a toroidal plasma current is generated in the tokamak via the transformer action technique. This toroidal current induces poloidal magnetic field, which generates a twist in the magnetic field lines. Following the twisted field lines, the charged particles move through the entire poloidal cross-section, thus generating a "shortcut" between the separated charges and avoiding the electric drift expelling the plasma.

2.3.1 TRAPPED AND PASSING PARTICLE ORBITS

Charged particle motion in tokamaks can be classified by two main groups: particles *passing* around the torus and particles *trapped* in magnetic wells formed by the difference in the magnetic field

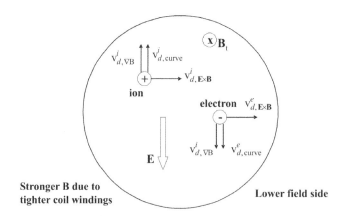

FIGURE 2.4 Guiding centre ∇B and curvature drift velocities of electrons and ions in a toroidal magnetic configuration (no poloidal magnetic field) result in vertical electric field and in outward electric drift of both electrons and ions.

strength between the inboard and outboard sides of the torus. The trapping of the particles depends on the ratio between particle velocity along the magnetic field and perpendicular to the magnetic field, and can be understood by considering (2.35) as the force on the magnetic moment μ of the particle orbit. The magnitude of the magnetic field along a particle trajectory in the torus has its lowest value, $B_{min} = B_0 \left(1 - \dfrac{r}{R_0} \right)$, in the median plane. Denoting the velocity at this point by a subscript zero and using the condition that μ is almost constant during the particle motion, we obtain

$$\frac{v_{\perp}^2}{B} = \frac{v_{\perp 0}^2}{B_{min}}. \tag{2.41}$$

If there is a bounce point along the particle orbit, $v_{\parallel} = 0$, the energy conservation implies that the perpendicular velocity at the bounce point satisfies $v_{\perp}^2 = v_{\perp 0}^2 + v_{\parallel 0}^2$. Substituting this expression into (2.41) gives the value of the magnetic field at the bounce point as

$$B_{bounce} = B_{min} \left[1 + \frac{v_{\parallel 0}^2}{v_{\perp 0}^2} \right]. \tag{2.42}$$

Thus, the pitch-angle $\dfrac{v_{\parallel 0}}{v_{\perp 0}}$ determines whether the particle is passing or trapped, as well as how short is the trajectory of the trapped particle between bounces. The boundary between trapped and passing particles is determined by the condition that the highest value of the magnetic field on the inner side of the torus is equal to the value of the "bounce" magnetic field,

$$B_{bounce} = B_{max} \approx B_0 \left(1 + \frac{r}{R_0} \right). \tag{2.43}$$

Thus, using (2.41), the requirement for the particle trapping, $B_{bounce} < B_{max}$, is

$$\frac{v_{\parallel 0}^2}{v_{\perp 0}^2} < \frac{2r}{R_0 - r}.$$

Using this critical condition for an isotropic distribution of particles, the fraction of the trapped particles as a function of minor radius r can be found to be approximately

$$n_{\text{trapped}} / n_{\text{total}} \approx \sqrt{2r/(R_0 + r)}. \tag{2.44}$$

The bounce orbits can be calculated using (2.35) for strongly trapped particles, which have their bounce points in the plasma cross-section (at some toroidal angle) within a narrow poloidal angle, $\vartheta \ll 1$, and $\dfrac{r}{R_0} \ll 1$. By writing the major radius coordinate

$$R = R_0 + r\cos\vartheta, \tag{2.45}$$

so that the magnetic field takes the form

$$B = \frac{B_0 R_0}{R} = \frac{B_0}{1 + \left(\dfrac{r}{R_0}\right)\cos\vartheta}, \tag{2.46}$$

and its parallel gradient, dB / ds is

$$\frac{\mathrm{d}B}{\mathrm{d}s} = \frac{rB_0}{R_0} \frac{\mathrm{d}\left(\dfrac{\vartheta^2}{2}\right)}{\mathrm{d}s}. \tag{2.47}$$

The equation of the field line is $\dfrac{r\mathrm{d}\vartheta}{\mathrm{d}s} = \dfrac{B_\vartheta}{B}$ so that $\vartheta = \dfrac{sB_\vartheta}{rB}$. Thus, combining (2.47) and (2.35), the equation of motion takes the form

$$\frac{\mathrm{d}^2 s}{\mathrm{d}t^2} = -\omega_b^2 s \tag{2.48}$$

where the bounce frequency of a trapped ion between the bounce points is

$$\omega_b = \frac{v_\perp}{qR_0} \cdot \sqrt{\frac{r}{2R_0}} \tag{2.49}$$

and the so-called "safety factor" is $q = \dfrac{rB_0}{R_0 B_\vartheta}$. We multiply both sides of (2.48) by ds/dt and integrate to obtain the equation of the particle motion along the magnetic field

$$s = s_b \sin(\omega_b t) \tag{2.50}$$

Because $\vartheta \propto s$, the ϑ-component of motion is given by

$$\vartheta = \vartheta_b \sin(\omega_b t) \tag{2.51}$$

and the equation of motion along the poloidal angle is

$$\frac{\mathrm{d}\vartheta}{\mathrm{d}t} = \omega_b \vartheta_b \tag{2.52}$$

The drift surface on which trapped particle orbit lies could be obtained by considering the radial component of the vertical drift due to the toroidal field gradient, $v_d = \frac{1}{2}m_j^2 v_\perp^2 \big/ \left(e_j B_t R\right)$, according to the following equation

$$\frac{\mathrm{d}r}{\mathrm{d}t} = v_d \sin\vartheta \approx v_d\vartheta. \tag{2.53}$$

By combining (2.52) and (2.53), we find

$$\frac{\mathrm{d}r}{\mathrm{d}\vartheta} = \frac{v_d}{\omega_b \vartheta_b} \cdot \frac{\vartheta}{\sqrt{1 - \dfrac{\vartheta^2}{\vartheta_b^2}}} \tag{2.54}$$

By integrating (2.54) we find the equation for the drift surface:

$$\left(r - r_0\right)^2 = \frac{\vartheta_b v_d}{\omega_b} \cdot \left(1 - \frac{\vartheta^2}{\vartheta_b^2}\right). \tag{2.55}$$

This surface has the shape of a banana, as shown in Figure 2.5, and the half-width of the drift orbit is $\Delta_O = \dfrac{\vartheta_b v_d}{\omega_b}$.

The guiding centre drift orbits of passing particles form trajectories similar to the magnetic flux surfaces, but shifted from the flux surfaces by the drift orbit size, as shown in Figure 2.5. Some regions of the particle phase space also form "non-standard" orbits, such as stagnation and potato orbits (not shown in Figure 2.5).

2.3.2   EXAMPLES OF FAT AND NON-STANDARD ORBITS OF HIGHLY ENERGETIC IONS IN JET

For highly energetic charged fusion products, such as fusion-generated alpha particles, the particle drift orbits may not be thin banana orbits, but fat orbits comparable to the minor radius of

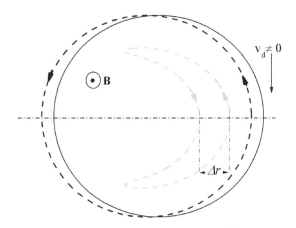

FIGURE 2.5   Guiding centre drift orbits of charged particles in a tokamak: trapped particle orbit ("banana" broken line) and passing particle orbit (circle broken line) shifted from the magnetic flux surface (circle line).

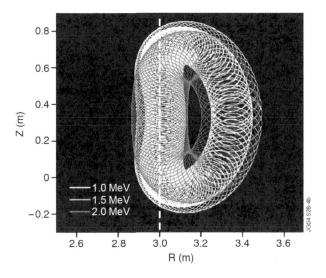

FIGURE 2.6 Examples of trapped orbits of He ions in the MeV energy range in JET tokamak.

the plasma. Figure 2.6 illustrates this point by showing orbits of trapped He ions accelerated with ICRH on JET up to the MeV energy range (experiment mimicking trapped fusion alpha particles). Furthermore, one can see here that Larmor radii are comparable to the drift orbits, and so full orbit description may be essential in solving some problems, such as interpreting fast ion losses to the first wall.

Next, some tokamak scenarios develop magnetic field topology significantly different than the one considered for discharges with toroidal current peaked at the plasma centre. To underline the role of the "twist" in magnetic field lines in the formation of drift particle orbits, we illustrate how energetic particle orbits appear in JET with strongly reversed magnetic shear equilibria, described in more detail in Chapters 4, 9, and 10. The reversed magnetic shear is formed in JET if the plasma current profile is hollow, that is, significantly reduced near the plasma centre, sometimes down to the zero value (so the safety factor $q(r) \to \infty$ in the plasma core). Such magnetic configurations are of interest for "advanced tokamak" scenarios aiming at triggering internal transport barriers. Fast ion distribution function resulting from on-axis ICRH satisfies

$$\Lambda = \frac{\mu B_0}{E} = 1. \tag{2.56}$$

The trajectories of hydrogen ions launched at the same energy 500 keV in the reversed-shear equilibrium (pulse #49382 discussed in more detail in Chapter 9) are shown in Figure 2.7. This distribution function consists of some trapped particles with bounce points, $V_\parallel = 0$ at the B contour through the magnetic axis, $B = B_0$. However, it also consists of non-standard ion orbits, for which $V_\parallel$ does not change sign. Because these orbits never bounce, that is, $\phi > 0$, its toroidal drift frequency is larger than the poloidal frequency. We see that many ions within q_{min} have orbits of the non-standard type, so some unusual equilibria could result in non-standard orbits deviating significantly from the usual trapped and passing types and should be inspected individually.

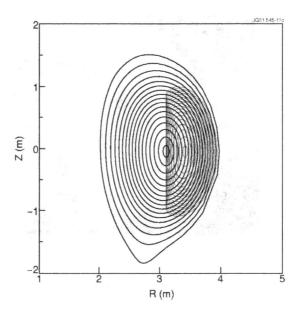

FIGURE 2.7   Drift orbits of ICRH-accelerated H-minority ions in JET pulse #49382 with reversed magnetic shear, $q_{min}=3.95$, $q(0)=8.5$. Twenty particles are launched between radii corresponding to the normalised poloidal magnetic fluxes $\psi=0.01$ and $\psi=0.99$ at the outboard mid-plane, $\vartheta=0$. The energetic ions have energy 500 keV and pitch-angles calculated such that $\Lambda = \mu B_0/E = 1$ corresponding to on-axis ICRH.

3 Energetic Ions in Present-Day Tokamaks

3.1 HEATING PLASMA WITH ENERGETIC IONS

After ionisation and starting the plasma, heating must be applied for achieving fusion-grade plasma temperatures to increase fusion reactivity, and alpha particle heating starts playing its role. At the early phase, tokamaks are heated with plasma current I_P induced by the transformer action, with Ohmic power P_{Ohm} determined by the plasma current and plasma resistivity R,

$$P_{\mathrm{Ohm}} = I_P V = \left[I_P \right]^2 R, \tag{3.1}$$

At low plasma temperatures, this Ohmic heating is very strong. However, the plasma resistivity depends on electron temperature T_e as

$$R \sim \left[T_e \right]^{-3/2}, \tag{3.2}$$

so when the plasma becomes hotter, its resistivity becomes smaller, and the Ohmic power becomes insufficiently effective. There is a maximum plasma temperature of ~5 keV, which can be achieved by Ohmic heating on present-day large tokamaks; however, auxiliary heating is needed to obtain ~10–20 keV temperature of plasma ions, which is optimal for D-T fusion.

One of the most effective and well-tested techniques of auxiliary heating of plasma up to ~10–20 keV temperatures is the heating via Coulomb collisions between thermal plasma species (electrons and plasma ions) and a population of low-density, $n_H \ll n_e$, but highly energetic (hot), $E_H \gg T_e$, T_i, ions produced in the plasma by auxiliary heating techniques (see Ref. [3.1] and References therein). Although the density of such energetic (hot) ions is small compared to the density of thermal plasma, the energy content of these energetic ions, $n_H E_H$, is not necessarily small in comparison to the thermal plasma energy content, $n_e T_e + n_i T_i$.

The hot ions of different energies transfer their energy to thermal ions and electrons at different proportions via Coulomb collisions. If the energy of hot ions is less than a critical value determined by

$$E_{\mathrm{crit}} = 14.8 A_f T_e \left(\sum_i n_i Z_i^2 \Big/ n_e A_i \right)^{2/3}, \tag{3.3}$$

the power of the hot ions flows mainly to thermal ions rather than to electrons. If the energy of hot ions is higher than the critical value (3.3), the hot ions mostly heat electrons. Here, A_f and A_i are the masses of fast ions and thermal ions, respectively, T_e is electron temperature, n_i and n_e are densities of thermal ions and electrons, respectively, and Z_i is the atomic number of thermal ions. The amount of energy going from hot ions with initial energy E into plasma thermal ions is given by the Stix formula [3.2]

$$G_i = \frac{E_{\mathrm{crit}}}{E} \int\limits_0^{E/E_{\mathrm{crit}}} \frac{\mathrm{d}y}{1 + y^{3/2}}, \tag{3.4}$$

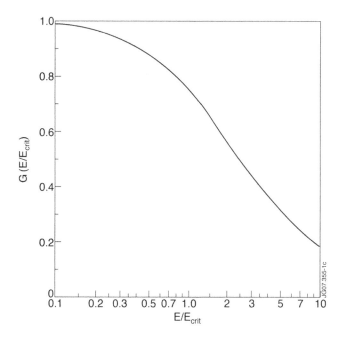

FIGURE 3.1   Graph of the function $G\left(E/E_{\mathrm{crit}}\right)$ given by the Stix formula (3.4).

Figure 3.1 illustrates the function $G_i\left(E/E_{\mathrm{crit}}\right)$.

The value of the critical energy implies that the largest ever existing tokamaks, JET, TFTR, JT-60U, DIII-D, and ASDEX-Upgrade, with neutral beam injection (NBI) as the main auxiliary heating technique have dominant ion heating of plasmas. This conclusion results from (3.3) taking into account the typical NBI energy range of ~ 60–160 keV and typical electron temperatures of ~5–10 keV. On the contrary, alpha particles with the energy of 3.52 MeV are well above the critical energy (3.3), and hence provide dominant electron heating. Moreover, NBI on small machines with low electron temperatures most often produce dominant electron heating.

The characteristic slowing-down time of energetic ions due to collisions with plasma electrons was first calculated by Spitzer [3.3]:

$$\tau_{\mathrm{se}} = 6.3 \cdot 10^{14}\, \frac{A_H T_e^{3/2}}{Z_H^2 n_e \ln\Lambda_e}, \tag{3.5}$$

where A_H, Z_H are the mass (in hydrogen mass units) and the charge number of the energetic ions, respectively, $n_e(\mathrm{m}^{-3})$ is the electron density, and $\ln\Lambda_e \approx 16$ is the Coulomb logarithm. The dependence of the slowing-down time on the mass of the energetic ions plays a key role in the experiment. For example, for the same plasma parameters, the slowing-down times of hydrogen, deuterium, and tritium energetic ions differ up to a factor of 3, implying that the tritium ions remain energetic for a much longer time than hydrogen or deuterium ions.

More details on the collisional relaxation of energetic ions are given in Appendix A. A comprehensive review validating the theory of energetic ion distributions in tokamak plasmas could be found in Ref. [3.4].

We now discuss the main mechanisms of plasma heating with energetic ions in tokamaks.

3.1.1 Neutral Beam Injection (NBI)

Among all techniques of auxiliary plasma heating, NBI plays a major role in present-day machines. In particular, all scenarios with high fusion performance rely on NBI-produced energetic ion populations as the principal source of plasma heating. Moreover, NBI is expected to play a key role in plasma heating and current drive in ITER.

In the NBI technique, deuterium ions from an ion source are accelerated via grids to a high energy. Then, they pass through the neutraliser and become neutral high energy atoms. Then, the highly energetic neutral beam is injected into plasma, with the penetration length of the atomic beam that depends on the NBI energy, mass, and the plasma density. Within the plasma, the NBI neutrals are ionised by collisions with thermal electrons and ions, and the resulting energetic ions are trapped by the magnetic field of the machine. Then, the energetic ion beam relaxation begins due to Coulomb collisions with thermal ions and electrons, leading to a steady-state beam distribution of the slowing-down type in energy,

$$f(v, \kappa) = \frac{3\beta_H B^2}{16\pi^2 m_H V_0^2} \cdot \frac{\Theta(V_o - v)}{v^3 + v_{\text{crit}}^3} h(\kappa), \tag{3.6}$$

where $\Theta(x)$ is the step function, V_o is the injection velocity of energetic ions, and $h(\kappa)$ with $\kappa \equiv v_\parallel / v$ is the pitch angle distribution. For an isotropic distribution of, for example, fusion-generated alpha particles, we have $h(\kappa) = 1$. The distribution function (3.6) is normalised to the average beta of the energetic ions:

$$\beta_H = \frac{8\pi m_H}{3B^2} \cdot \int f \cdot v^2 d^3 v.$$

Sometimes, the distribution function of NBI-produced energetic ions deviates from the slowing-down form (3.6) significantly. This happens when mechanisms other than Coulomb collisions affect the beam ions on a time scale shorter than the slowing-down time. For example, if the characteristic time of charge-exchange of the beam ions is shorter than the beam slowing-down time, the beam with a constant source can have a steady-state distribution with a bump-on-tail in the energy.

Apart from deuterium, beams used in present-day machines can inject hydrogen, tritium, He, and He^3. Hydrogen gases produce neutral beams in three fractions at E, $E/2$, and $E/3$ energies, whereas helium beams are produced at energy E only. In addition to plasma heating, NBI also provides fuelling. In particular, tritium NBI at an energy of 160 keV was very effective in penetrating to the plasma core of JET tokamak, thus providing tritium fuelling close to the optimum D:T = 50:50 mixture in high fusion power D-T experiments in JET [3.5].

Let us now summarise the advantages and disadvantages of NBI.

Advantages of NBI
Efficient heating of thermal ions in present-day experiments;
High power capability (40 MW on TFTR, >30 MW on JET);
Drives plasma rotation (thus affecting MHD instabilities, i.e., stabilising lock modes);
Provides fuelling;
Provides current drive;
Not sensitive to the value of magnetic field.

Disadvantages
Needs MeV energy beams for penetrating in ITER $\rightarrow$ Negative ion source for NBI is needed;
Heating not well localised;
Large aperture.

3.1.2 Ion Cyclotron Resonance Heating (ICRH)

ICRH is used for creating a highly energetic population of some selected ions (a "tail" in the distribution function of these ions), which could further deliver heating to both electrons and thermal ions depending on the ratio between the critical energy and the tail temperature [3.6]. In contrast to NBI where the beam energy is limited by the energy of the NBI source, ICRH could accelerate ions up to the MeV range. In some cases, ICRH-accelerated helium ions can mimic fusion-generated alpha particles for important applications, such as developing and testing alpha particle diagnostics in a radiation-free plasma environment [3.7].

In the ICRH technique, a fast magneto-acoustic wave is launched by an external antenna with the frequency of the ion cyclotron range, which could be in resonance with the selected ions. A low field side antenna is usually used, as shown in Figure 3.2. The radio-frequency (RF) power density absorbed by the ions with mass m_H and distribution function f resonating via $\omega = n \cdot \omega_{BH}$ with the wave launched is given by [3.8]:

$$p_{\text{absorbed}} = -\int m_H V_\perp^2 D_{\text{RF}} \frac{\partial f}{\partial V_\perp} dV_\perp \tag{3.7}$$

where RF diffusion is given by

$$D_{\text{RF}} \propto \left(E_+ J_{n-1}\left(\frac{k_\perp V_\perp}{\omega_{\text{BH}}} \right) + E_- J_{n+1}\left(\frac{k_\perp V_\perp}{\omega_{\text{BH}}} \right) \right)^2, \tag{3.8}$$

$V_\perp$ is the ion velocity perpendicular to the magnetic field, ω_{BH} is the ion cyclotron frequency, $k_\perp$ is the perpendicular wave-number of the fast wave launched, and $J_{n-1}(x) J_{n+1}(x)$ are the Bessel functions of first kind.

The wave launched by the antenna is an elliptically polarised mode and its electric field can be decomposed as a sum of two components: the left-hand polarised component E_+, which rotates in the ion direction, and the right-hand polarised component E_-, which counter-rotates. For $n \geq 2$, the relation between these two components is approximately given by

$$\frac{E_-}{E_+} = \frac{n+1}{n-1}. \tag{3.9}$$

Among the Bessel functions, only Bessel function of the zeroth order has a finite value at zero argument, $J_0(0) \neq 0$. This case corresponds to $n = 1$ fundamental cyclotron resonance,

$$\omega = \omega_{\text{BH}}. \tag{3.10}$$

The best known and most widely explored hydrogen minority ICRH [3.6], which accelerates low-density H-minority population right from the thermal energy, is of the type (3.10). For higher-order resonances, $n > 1$, the argument of the Bessel functions in (3.8) must be non-zero. For example, in JET experiment with ICRH acceleration via third deuterium cyclotron harmonic, a "seed" energetic D beam with high $V_\perp$ obtained from NBI at ~100 keV worked well [3.9].

For beam ions with high parallel velocity, the cyclotron resonance has a Doppler shift, the value of which could be significant,

$$\omega = \omega_{\text{BH}} + k_\parallel v_\parallel. \tag{3.11}$$

In this case, the ions are accelerated at major radius shifted from the vertical line given by (3.10) and shown in Figure 3.2. Moreover, the topology of the drift ion orbits accelerated in such a manner can significantly differ from the usual trapped banana orbits.

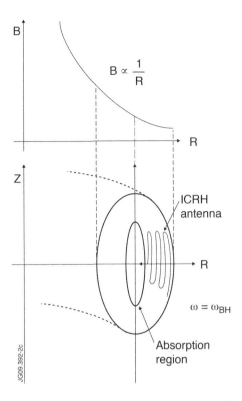

FIGURE 3.2   Schematic representation of ICRH technique used in on-axis H-minority heating.

Figure 3.2 illustrates the geometry of ICRH for the on-axis H-minority case (3.10). In a toroidal solenoid, magnetic field at the outer side is lower than that at the inner side, and the radial gradient of the equilibrium magnetic field, $B(R) \sim B_0/R$ makes the cyclotron frequency a function of major radius too. Because the wave with frequency ω propagating from the antenna matches the local cyclotron frequency of H-minority ions, $\omega_{BH}(R)$, at some point inside the plasma, the cyclotron resonance $\omega = \omega_{BH}(R)$ becomes possible at this point. During the resonant interaction between the wave and the H-minority ions, an exchange of energy from the wave to the ions increases mostly perpendicular energy of the ions, which is further delivered to thermal ions and electrons via Coulomb collisions.

The advantages and disadvantages of ICRH could be summarised as follows:

Advantages of ICRH
Localised heating;
The tail temperature could be in the MeV range;
Hydrogen minority ICRH creates H-minority with $E > E_{crit}$ – it heats electrons;
Heating of thermal ions is also possible, for example, with ^{3}He minority in DT plasma;
Some current drive.

Disadvantages
Antenna inside the vessel;
Relatively low power capability;
Plasma coupling may be a problem in the presence of edge-localised modes (ELMs), for example;
Power deposition area depends on the magnetic field.

3.1.3 ALPHA PARTICLE HEATING AND BURNING PLASMAS

Alpha particle heating in ignited plasmas will dramatically change the plasma scenario as the plasma will become an exothermic and highly non-linear medium. To approach such self-heated ignited fusion plasma, a concept of burning plasma was introduced, within which some auxiliary heating is still used for plasma control, in addition to plasma self-heating by fusion alpha particles.

The effects associated with alpha particles become increasingly significant as the ratio between the alpha particle heating power, $P_\alpha = 0.2\, P_{FUS}$, and the auxiliary heating power, P_{IN}, increases:

$Q \equiv P_{FUS}/P_{IN} \approx 1$ – at the threshold;
$Q \approx 5$ – alpha particles significantly affect plasma heating, and may excite AEs;
$Q \approx 10$ – non-linear coupling becomes important between alpha particles, MHD events, turbulent transport, and interaction plasma-boundary;
$Q \geq 20$ – dominant self-heating and transient ignition;
$Q \to \infty$ – ignition.

With all the above-mentioned effects, the problem of predicting with confidence how to maximise the alpha particle heating and, simultaneously, minimise alpha particle losses to the first wall becomes very challenging. An extended study of fast ions is being performed for solving this problem along the following directions: (1) experiments in present-day machines with fast ions produced by ICRH and NBI, which could match the ITER key dimension-less parameters $\beta_{fast}/\beta_{thermal}$, V_{fast}/V_A, $R_0 \cdot d(\beta_{fast})/dr$; (2) development of diagnostics of fast ions; and (3) theoretical analyses and numerical simulations of fast ion phenomena observed in present-day experiments, with further extrapolation to burning plasmas.

Currently, a source of the largest uncertainty in the predictions of alpha particles in burning plasmas is the possible excitation of Alfvén instabilities via resonance interaction with alpha particles. The problem is in the very high energy of 3.52 MeV of alpha particles, which makes these fusion-generated ions *super-Alfvénic*,

$$V_{Ti} \sim 10^6\,\text{m/s} \ll V_A \sim 5 \times 10^6\,\text{m/s} < V_\alpha = 1.3 \times 10^7\,\text{m/s} \ll V_{Te} \sim 8 \times 10^7\,\text{m/s}, \quad (3.12)$$

where V_A is Alfvén speed, and the estimates are for D:T=50:50 plasmas of the ITER machine with $B_T = 5.3$ T. During the slowing-down process, the alpha particles pass the resonance condition

$$V_A = V_{\|\alpha} \quad (3.13)$$

and may excite Alfvén waves. The free energy source for such instability is associated with the radial gradient of alpha particle pressure. The instability results in a radial re-distribution of alpha-particles changing the power deposition profile and possibly causing losses of highly energetic alphas.

3.2 DIAGNOSTICS OF CONFINED ENERGETIC IONS

Diagnostics of energetic ions and energetic ion-driven instabilities are crucial for burning plasmas. Development of such diagnostics is challenging as the diagnostics need to perform accurate simultaneous measurements of several populations of energetic ions, confined and lost, under the harsh conditions of D-T plasma. In addition, reliable techniques for detecting Alfvénic instabilities coupled to the energetic ions via wave-particle resonances are required as these instabilities may affect the transport and losses of energetic ions.

This section reviews the diagnostic techniques for energetic ions employed on tokamaks.

3.2.1 MEASUREMENTS OF ENERGETIC ION DISTRIBUTION WITH A NEUTRAL PARTICLE ANALYSER

Measurements of confined energetic ions are often carried out by analysing flux of neutrals escaping from plasma. The well-known instrument for this is the neutral particle analyser (NPA), which relies on the neutralisation of the energetic ions in charge-exchange reactions with some electron donors in the plasma. When the electron donors are naturally present in the plasma, for example, carbon or beryllium impurities, the NPA measurements are passive. If a source of electron donors is used, for example, NBI, then the measurements are active. The escaping neutrals are re-ionised in gas cells or with stripping foils at the entrance to NPA, and deflected by magnetic and electric fields to determine their energy and mass. The NPA measurements are intended: (1) to provide experimental testing and validation of theories of ICRH and NBI heating scenarios, (2) to quantify the link between energetic ions and MHD instabilities, and (3) to investigate confinement and slowing down of charged fusion products.

For example, consider NPA for high energy, $0.3 \leq E$ (MeV) ≤ 3.5, hydrogen and helium isotope fluxes [3.10] on JET. This NPA is located at the top of the torus with its vertical line-of-sight intersecting the ICRH power deposition region and lower energy (Octant 4) NBI at the plasma centre. The energy distribution function $f(E)$ integrated along the line-of-sight is measured for ions with pitch angle $\leq 5 \times 10^{-3}$.

Procedures for the deduction of the local effective temperature and minority density in the case of an anisotropic distribution function of ICRH-accelerated ions from the measured $f(E)$ were developed for NPA in such geometry in Ref. [3.11]. The NPA measurements of the energy distribution functions of ICRH-accelerated ions have resulted in corroboration of a linear increase in the perpendicular temperature of the minority ions with the ICRH power density [3.11], as predicted by the Stix model [3.5]. Typical NPA measurements are shown in Figure 3.3 for JET discharge, in which both H-minority ICRH and D-NBI were used [3.12]. Figure 3.3 demonstrates a reduction in the tail temperature of the H-minority distribution function from 243 to 160 keV, after deuterium NBI with starting energy of 140 keV starts at 12 s. Starting from that time, a high energy tail builds up to

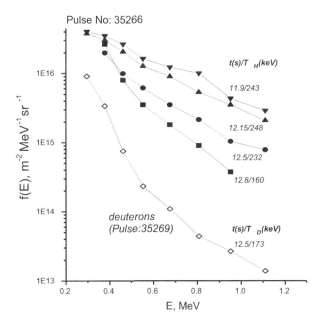

FIGURE 3.3 Evolution of ICRH-accelerated proton distribution function during combined ICRH and deuterium NBI. Distribution function of the second harmonic ICRH-driven beam deuterons is also shown.

173 keV in the deuterium energy distribution, implying a channelling of the ICRH power from the first harmonic heating of H-minority to the second harmonic ICRH of the D beam.

Measurements of fusion alpha-particles were made with the NPA during JET D-T campaign in 1997 [3.13], in a high fusion yield H-mode discharges with NBI heating only, comprising 80 keV D atoms and 140 keV T atoms, giving a D-T fuel mixture of $n_D/(n_{D+}n_T)\sim0.5$. The measurements were carried out with the NPA which was set up to detect ^{4}He atoms in eight channels in the energy range 0.8–3.0 MeV. For the DT plasma analysed, two surprising findings were observed: (1) the flux of neutrals corresponding to the helium mass-to-charge ratio was an order of magnitude higher than expected, and (2) the low energy flux of neutrals corresponding to helium was seen almost immediately after NBI power was switched on, that is, well before the fusion-generated alpha particles could slow down to the low energy range of the NPA measurements. The analysis [3.13] has identified the excessive flux as that of high energy deuterons accelerated by close elastic collisions (knock-on) between fusion alpha particles and thermal D ions [3.14]. Despite the low density of knock-on deuterons compared to that of alpha particles ($0.0025 \le f_d(E)/f_\alpha(E) \le 0.07$), a flux of deuterium atoms to the low-energy NPA channels exceeding the helium flux could be produced because of the much higher (about a factor of 100) neutralisation probability for the single-charged deuterium ions than for double-charged alpha-particles. The measured distribution functions of the helium-like ions [3.13], were later shown to be in a satisfactory agreement with the Fokker–Planck code FPP-3D, which incorporates neoclassical transport and classical slowing down of fusion alpha-particles and the knock-on deuterons [3.15]. Figure 3.4 illustrates the comparison.

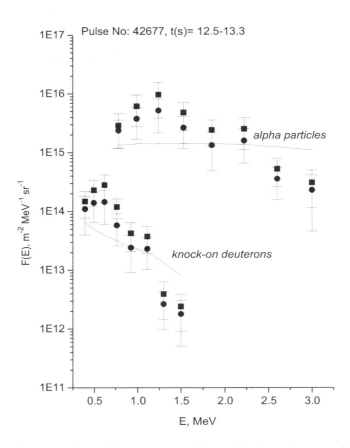

FIGURE 3.4 Distribution functions of fusion alpha particles and knock-on deuterons: deduced from NPA measurements (points) and modelling assuming classical confinement and slowing down (curves). Circles and squares show the results when the maximum and minimum values for neutralisation probability are used.

3.2.2 GAMMA RAY DIAGNOSTICS OF ENERGETIC IONS

One of the most interesting diagnostics of energetic ions with significant progress in recent years is the diagnostic of energetic ions via nuclear gamma rays [3.16]. A strong gamma ray emission in the MeV energy range comes from fusion-grade tokamak plasmas. This emission has both a continuum spectrum in energy and a discrete spectrum. The continuum spectrum is caused by the Bremsstrahlung, supra-thermal electrons, and a background noise, whereas the discrete spectrum consists of spectral lines emitted in nuclear reactions between energetic and thermal ions and Be and C impurities. Each type of energetic ions colliding emits a well-determined gamma ray with energy unique to this specific ion colliding with a specific impurity (as fingertips are unique to a certain person).

For ITER, a simultaneous diagnosis of two types of fast ions, the fusion-generated alpha particles (at 3.52 MeV) and NBI-produced deuterium ions (born at about 1 MeV), will be of major importance as these will be the fast ion populations with the largest energy contents. In the presence of Be impurity, energetic ^{4}He ion with $E > 1.7$ MeV could be detected from gamma rays with an energy above 4 MeV arising from the nuclear reactions

$$^4\text{He}(E > 1.7 \text{ MeV}) + {^9}\text{Be} \rightarrow {^{12}}\text{C} + \text{n} + \gamma, \tag{3.14}$$

while energetic deuterium ions with $E > 0.5$ MeV could be detected from gamma rays generated during the reaction

$$\text{d}\,(E > 0.5 \text{ MeV}) + {^{12}}\text{C} \rightarrow {^{13}}\text{C} + \text{p} + \gamma. \tag{3.15}$$

Figure 3.5 shows the spectrum of gamma rays as a function of gamma ray energy measured in one of JET discharges. In this figure, the double peak around 4 MeV results from nuclear reaction (3.14),

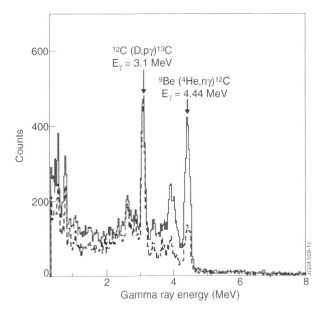

FIGURE 3.5 Gamma ray spectra measured by the NaI(Tl) detector on JET: solid line shows the spectrum recorded in JET discharge with 70 and 110 keV He NBI; dashed line shows the spectrum recorded in a discharge with two 70 keV He NBI.

whereas the double peak around 3 MeV results from nuclear reaction (3.15). The significant energy separation between these two peaks was used in measuring gamma emission with a 19-channel two-dimensional (2D) camera on JET, as shown in Figure 3.6. Namely, each channel of this 2D camera was prescribed to measure simultaneously gamma rays in two different energy bands, one of which corresponds to the gamma rays coming from the reaction (3.14) of fast ^{4}He and the other to the gamma rays (3.15) from fast D. Hence, the spatial profiles of gamma rays from two ITER-relevant fast ion species were measured simultaneously for the first time, as shown in Figures 3.7 and 3.8, with the integration time of 1 s.

In these alpha particle simulation experiments [3.7] with third harmonic ICRH of ^{4}He-beam in ^{4}He-plasmas, gamma-radiation due to the nuclear reaction of ^{9}Be(^{4}He, nγ)^{12}C was detected [3.17], showing the successful ICRH acceleration of ^{4}He-beam ions. The first energy level 4.44 MeV of the final nucleus, ^{12}C, is excited by alpha particles resulting in a peak at 4.44 MeV. The gamma ray emission from the reaction ^{12}C(D, pγ)^{13}C was observed as well. A peak at 3.09 MeV (transition 3.09 → 0), which is identified as a gamma emission from the ^{12}C(D, pγ)^{13}C reaction, reflects the presence in the plasma of fast D ions in the MeV range. This indicates that the residual deuterium minority in helium plasmas also absorbs some ICRH power at the third harmonic D resonance that coincides with the third harmonic ^{4}He resonance.

This gamma ray diagnostic technique is especially important for burning plasmas where several groups of different energetic ions co-exist, in addition to the alpha particles, to control the burn. Further development of gamma ray diagnostics will inevitably result in the compatibility of gamma ray diagnostics with DT operation. This is possible with the use of LiH neutron filters transparent to gamma rays, but must be tested in DT plasma first.

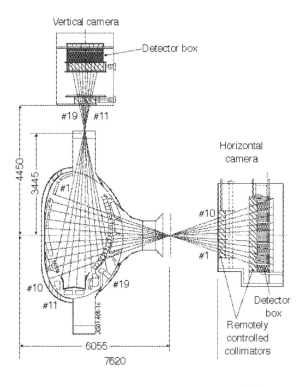

FIGURE 3.6   Schematic of the two-dimensional gamma ray camera on JET.

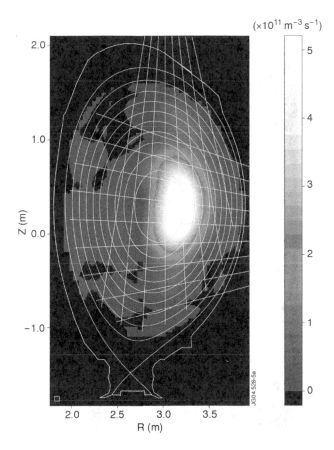

FIGURE 3.7   Two-dimensional tomographic reconstruction of spatial profile of the gamma rays from fast D ions.

3.2.3   Information on Highly Energetic D Ions in the Neutron Emission Spectra

The effect of D beam acceleration with third harmonic ICRH in D plasma [3.9] gives in JET a spectacular increase in D-D fusion yield and a significant increase in the energy of the generated D-D neutrons. The effect of the D-D neutron energy increase was studied in-depth on JET with the time-of-flight TOFOR spectrometer [3.18]. The TOFOR device is placed well above the JET machine, with the line-of-sight passing through the magnetic axis perpendicular to the magnetic field. Because ICRH mostly increases the beam ion velocity perpendicular to the magnetic field, the interpretation of the D-D neutron spectrum does not involve the beam velocity parallel to the magnetic field, and is relatively straightforward. In particular, it is possible to establish a correspondence between the maximum energy of the projectile D ions and the maximum energy of D-D neutrons resulting from the fusion of these projectile ions and thermal D plasma [3.19]. Figure 3.9 shows an example of JET discharge (pulse # 86459 with B_T=2.26 T, I_P=2.16 MA), with the combined synergetic NBI and third beam harmonic ICRH, with the TOFOR measurements of D-D neutrons made at different times.

As shown in Figure 3.9, by adding ~3 MW of ICRH power at third harmonic to ~4.5 MW of D NBI in D plasma, it was possible to increase the yield of D-D neutrons by a factor of ~10. Indeed, the TOFOR measurements show much lower yield of D-D neutrons during the NBI-only heating (see Figure 3.3a and b) phase than during the NBI+ICRH phase (see Figure 3.3c and d). Furthermore,

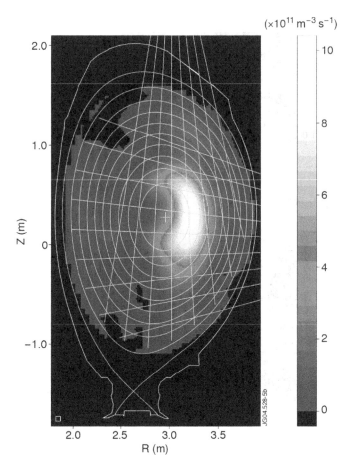

FIGURE 3.8   Two-dimensional reconstruction of spatial profile of the gamma rays from fast ^{4}He ions.

the fastest time-of-flight decreases significantly, in line with the expected broadening of the energy spectrum of the D-D neutrons. Note that the time-of-flight of D-D neutrons measured with TOFOR depends on the neutron energy as $\tau_{TOF} \propto E_{-n}^{-1/2}$ giving a value of ~65 ns for 2.5 MeV neutrons, and much shorter time of ~45 ns for 5 MeV neutrons.

Measurements of neutron emission spectrum were also used in D-T plasmas on JET in 1997. These provided information on the fusion reactivity in D and D-T plasmas, including its dependence on the velocity distribution of fuel ions. Thermal Maxwellian ions produce a Gaussian spectra, the width of which is determined by the Doppler broadening reflecting ion temperature. A deviation from the Gaussian shape indicates the presence of supra-thermal velocity components which appear in conjunction with ICRH and/or NBI, or with the high energy tail due to the knock-on effect [3.14, 3.19]. The spectrum of neutrons born in reactions with D and T knock-on ions extends in energy well beyond that of D-T neutron emission even with ICRH and NBI. The knock-on effect on the spectrum of D-T neutrons was observed experimentally for the first time during the DTE1 campaign on JET (1997) and was also theoretically modelled [3.20]. The knock-on tail was identified in the highest energy of 20% in the measured spectrum corresponding to the neutron energy range $E_n = 15.7–16.8$ MeV. The observation is well described by a calculation with respect to the knock-on neutron flux/thermal neutron flux ratio assuming that the alpha particle confinement and slowing down is classical.

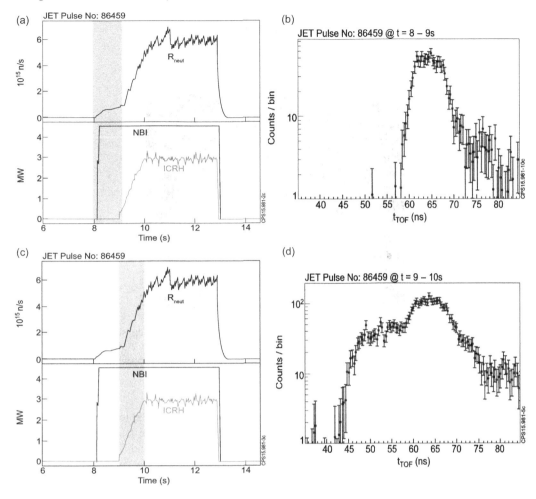

FIGURE 3.9 ICRH and NBI power wave-forms and time-of-flight of D-D neutrons measured in JET discharge # 86459. The early phase with NBI-only marked in grey in (a) generates D-D neutron yield at ~15 counts/bin with the fastest neutron time-of-flight of ~60 ns as (b) shows. Adding 3 MW of ICRH at the third harmonic of D beam marked in grey in (c) increases the yield of D-D neutrons to ~150 counts/bin with much faster neutron time-of-flight at ~45 ns as (d) shows.

3.3 DIAGNOSTICS OF LOST ENERGETIC IONS

Lost energetic ions are detected on many machines with a series of Faraday cups and/or scintillators.

The Faraday cups represent a set of several parallel foils placed inside the torus where the flux of lost energetic ions is investigated. In the presence of lost energetic ions, each of the foils provides counts of the ions hitting it, and as the foil number in the set hit by the lost ion depends on the ion-penetrating ability, the Faraday cups provide some resolution in the energy of the lost energetic ions. On JET, the Faraday cups [3.21] consist of an array of 13 individual detectors, each with a minimum of four foils. Each detector allows a modest degree of energy resolution (between 20% and 50% depending on the detailed foil and aperture geometry), a time resolution of about 1 ms, and a minimum detectable signal of about 0.5 nA. The detectors are distributed between three radial locations (equally spaced between 25 and 85 mm behind the adjacent poloidal limiter) at five poloidal locations between mid-plane and 80 cm below mid-plane. The detectors have an energy resolution between 30% and 50% and a bandwidth of 1 kHz. The Faraday cups diagnostic is compatible with the harsh D-T conditions, robust, and suitable for burning plasmas.

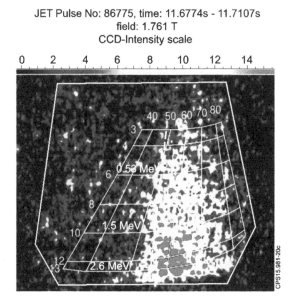

FIGURE 3.10 Scintillator probe data (grid of gyro-radius, cm, and pitch-angle, deg) showing losses of alpha particles born in D-^{3}He fusion reactions in JET discharge #86775 (B_T=2.24 T, I_P=2 MA). Larmor radii corresponding to alpha particles with energies of 0.53, 1.5, and 2.6 MeV are shown. The white broken line shows the separatrix at the trapped/passing boundary of alpha particles.

The Faraday cups on JET are complemented by a scintillator probe [3.22] located just below the equatorial plane. The scintillator probe consists of a thin layer of P56 scintillator (0.63 lm/W). The image of the scintillator surface is transmitted through a series of lenses and fibre-optic bundles to a CCD and a photomultiplier tube outside the vessel with respective time resolution of 20 and 3 ms. The device provides a pitch angle resolution of 5%, and a best achievable resolution of the gyro-radius of 15%. To maintain a scintillator temperature below approximately 400°C, an externally supplied water cooling system is incorporated in the probe design.

Figure 3.10 shows an example of alpha particle measurements with the scintillator probe on JET. The third harmonic ICRH acceleration of D beams in D-3He plasmas was employed in this experiment with amounts of ^{3}He increasing discharge-by-discharge of up to $n_{He}/n_e \approx 30\%$ [3.9]. The scintillator probe was employed for measuring ICRH-accelerated D ions and charged fusion products of both D-^{3}He and D-D reactions. Alpha particles generated via D-^{3}He fusion were seen in the scintillator probe as the losses with Larmor radii in excess of ~10 cm, just after both NBI and ICRH were switched off. Figure 3.10 shows the scintillator probe signal at the time of interest.

REFERENCES

1. J. Wesson, *Tokamaks*, Oxford University Press, 4th Edition (2011).
2. T.H. Stix, *Plasma Phys.* **14** (1972) 367.
3. L. Spitzer, *Physics of fully ionised gases*, Interscience, New York (1962).
4. W.W. Heidbrink and G.J. Sadler, *Nucl. Fusion* **34** (1994) 535.
5. M. Keilhacker et al., *Nucl. Fusion* **39** (1999) 209.
6. T.H. Stix, *Nucl. Fusion* **15** (1975) 737.
7. M.J. Mantsinen et al., *Phys. Rev. Lett.* **88** (2002) 105002.
8. L.-G. Eriksson et al., *Nucl. Fusion* **38** (1998) 265.
9. S.E. Sharapov et al., *Nucl. Fusion* **56** (2016) 112021.
10. A.I. Kislyakov et al., *Fusion Eng. Des.* **34–35** (1997) 107.
11. K.G. McClements et al., *Nucl. Fusion* **37** (1997) 473.

12. D. Testa et al., *Phys. Plasmas* **6** (1999) 3498.
13. A.A. Korotkov et al., *Phys. Plasmas* **7** (2000) 957.
14. D.D. Ryutov, *Phys. Scripta* **45** (1992) 153.
15. F.S. Zaitsev et al., *Nucl. Fusion* **42** (2002) 1340.
16. V.G. Kiptily et al., *Nucl. Fusion* **42** (2002) 999.
17. V.G. Kiptily et al., *Nucl. Fusion* **45** (2005) L21.
18. M. Gatu Johnson et al., *Nucl. Instrum. Methods* **A591** (2008) 417.
19. J. Eriksson et al., *Plasma Phys.* Control. Fusion **55** (2013) 015008.
20. J. Källne et al., *Phys. Rev. Lett.* **86** (2000) 1246.
21. D. Darrow et al., *Rev. Sci. Instr.* **75** (2004) 3566.
22. S. Baeumel et al., *Rev. Sci. Instr.* **75** (2004) 3563.

4 Equilibrium of Tokamak Plasma

4.1 GOVERNING IDEAL MHD EQUATIONS FOR TOKAMAK PLASMAS

For describing plasma, we take velocity moments of collision-less kinetic equations for distribution functions of electrons and thermal ions of the plasma and sum the results over the plasma species to obtain

$$\frac{\partial \rho}{\partial t} + \nabla \cdot (\rho V) = 0, \tag{4.1}$$

$$\rho \frac{dV}{dt} = -\nabla p + \frac{1}{c} J \times B, \tag{4.2}$$

$$\frac{\partial \rho}{\partial t} + V \cdot \nabla p + \gamma p \nabla \cdot V = 0, \tag{4.3}$$

Here (4.1) is the mass conservation equation, ρ is the plasma mass density, and V the fluid velocity. The equation of plasma motion is given by (4.2), where

$$\frac{d}{dt} = \frac{\partial}{\partial t} + V \cdot \nabla,$$

p is the plasma pressure, and J, B are the plasma current and the magnetic field, respectively. Adiabatic behaviour is assumed for plasma pressure described by (4.3), where γ is the adiabaticity index.

The perpendicular Ohm's law is obtained from the equation of electron motion,

$$E + \frac{1}{c} V \times B = 0, \tag{4.4}$$

Here, we assume a perfectly conducting plasma so that no electric field E can be sustained in the moving fluid reference frame as Eq. (4.4) shows.

Although Eqs. (4.1)–(4.4) are obtained from the kinetic equations, they look similar to hydrodynamic equations describing current-carrying liquid in magnetic field; therefore, they are called ideal magneto-hydrodynamic (MHD) equations [4.1].

For describing electromagnetic fields in the plasma, Maxwell's equations are used, which in CGS units are:

$$\nabla \times B = \frac{4\pi}{c} J, \tag{4.5}$$

$$\nabla \times \boldsymbol{E} = -\frac{1}{c}\frac{\partial \boldsymbol{B}}{\partial t}, \tag{4.6}$$

$$\nabla \cdot \boldsymbol{B} = 0. \tag{4.7}$$

Scale lengths larger than Debye length are considered with the plasma quasi-neutrality condition,

$$n_e = \sum_i Z_i \, n_i \tag{4.8}$$

For further analysis, we express plasma current as the sum of the current components parallel and perpendicular to the magnetic field:

$$\boldsymbol{J} = \frac{(\boldsymbol{J} \cdot \boldsymbol{B})}{B^2}\boldsymbol{B} + \frac{\boldsymbol{B} \times (\boldsymbol{J} \times \boldsymbol{B})}{B^2}. \tag{4.9}$$

The second term is determined by the equation of motion (4.2):

$$\frac{1}{c}\boldsymbol{B} \times (\boldsymbol{J} \times \boldsymbol{B}) = \left(\boldsymbol{B} \times \rho\frac{d\boldsymbol{V}}{dt}\right) + \left(\boldsymbol{B} \times \nabla p\right) \tag{4.10}$$

Next, from Amperes law (4.5), we obtain

$$\nabla \cdot \boldsymbol{J} = \nabla \cdot \left(\frac{c}{4\pi}\nabla \times \boldsymbol{B}\right) = 0 \tag{4.11}$$

By substituting Eqs. (4.9) and (4.10) into (4.11) and using (4.7), we arrive at the main governing equation of ideal MHD

$$\boldsymbol{B} \cdot \nabla\left(\frac{\boldsymbol{J} \cdot \boldsymbol{B}}{B^2}\right) + \nabla \cdot \left(4\pi\rho\frac{\boldsymbol{B} \times d\boldsymbol{V}/dt}{B^2}\right) + \nabla \cdot \left(\frac{\boldsymbol{B} \times \nabla p}{B^2}\right) = 0. \tag{4.12}$$

We introduce equilibrium (subscript 0) and perturbed (denoted by δ) quantities:

$$\boldsymbol{J} = \boldsymbol{J}_0 + \delta\boldsymbol{J}, \boldsymbol{B} = \boldsymbol{B}_0 + \delta\boldsymbol{B}, \boldsymbol{E} = \delta\boldsymbol{E}, \boldsymbol{V} = \delta\boldsymbol{V},$$

$$\rho = \rho_0 + \delta\rho, p = p_0 + \delta p \tag{4.13}$$

A static equilibrium means $d/dt=0$, so we obtain the following for the static and axi-symmetric (independent of toroidal angle) equilibrium of tokamak plasma:

$$\nabla p_0 = \frac{1}{c}\boldsymbol{J}_0 \times \boldsymbol{B}_0. \tag{4.14}$$

Therefore, $\boldsymbol{B}_0 \nabla p_0 = 0$, $\boldsymbol{J}_0 \nabla p_0 = 0$, that is, the plasma pressure is constant along both the magnetic field lines and along the current lines; the plasma expands freely along these lines. Next, because ∇p_0 is perpendicular to the surface $p_0 = \text{const}$, the magnetic field lines and the current lines must lie on the $p_0 = \text{const}$ surfaces. In a tokamak, the magnetic field lines lie in nested toroidal magnetic surfaces, as shown in Figure 4.1 [4.2].

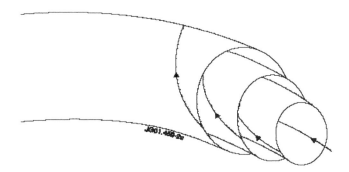

FIGURE 4.1   Schematic showing magnetic flux surfaces enclosed within each other in torus.

4.2   GRAD–SHAFRANOV EQUATION FOR TOKAMAK PLASMA EQUILIBRIUM

For assessing the equilibrium plasma equation (4.14), we introduce equilibrium poloidal magnetic field flux function ψ and equilibrium poloidal current density flux function:

$$\boldsymbol{B}_{\vartheta} = \frac{1}{R}[\nabla\psi\times\boldsymbol{e}_{\phi}]; \quad \boldsymbol{J}_{\vartheta} = \frac{2}{cR}[\nabla I\times\boldsymbol{e}_{\phi}] \tag{4.15}$$

The equilibrium equation (4.14) expressed via the poloidal and toroidal components of the current and magnetic field has the following form:

$$\boldsymbol{J}_{\vartheta}\times\boldsymbol{e}_{\phi}B_{\phi} + \boldsymbol{J}_{\phi}\times\boldsymbol{e}_{\phi}B_{\vartheta} = c\nabla p. \tag{4.16}$$

Here, we drop the subscript "0" in this section.

By substituting (4.15) into (4.16), we obtain the form of *Grad–Shafranov equation* [4.3, 4.4] for the poloidal flux ψ:

$$R^2\nabla\cdot\left(\frac{1}{R^2}\nabla\psi\right) + I\frac{dI}{d\psi} + 4\pi R^2\frac{dp}{d\psi} = 0. \tag{4.17}$$

For further analysis, we introduce a properly chosen magnetic field coordinate system (see Appendix B) and use the expressions for coordinates from this Appendix to obtain:

$$R^2\nabla\left(\frac{1}{R^2}\nabla\psi\right) = \frac{R^2}{J}\frac{\partial}{\partial r}\left(\frac{J}{R^2}\psi'g^{11}\right) + \frac{R^2}{J}\frac{\partial}{\partial\vartheta}\left(\frac{J}{R^2}\psi'g^{12}\right) = \frac{1}{r}\frac{\partial}{\partial r}(r\psi')$$

$$+\cos\vartheta\left(\frac{1}{r}\frac{\partial}{\partial r}(2\Delta'r\psi') - \frac{1}{r}\left(\frac{r}{R_0} + (r\Delta')'\right)\psi'\right);$$

$$I\frac{dI}{d\psi} = I\frac{dI}{dr}\left(\frac{d\psi}{dr}\right)^{-1};$$

$$4\pi R^2\frac{dp}{d\psi} = \left(\frac{4\pi R_0^2}{\psi'}\right)p'\left(1 + 2\frac{r}{R_0}\cos\vartheta\right). \tag{4.18}$$

Equations (4.17) and (4.18) could be analysed by combining the ϑ-independent terms to obtain:

$$\frac{1}{r}\left(r\psi'\right)' + \frac{II'}{\psi'} + \frac{4\pi R_0^2}{\psi'} p' = 0, \tag{4.19}$$

which could also be represented as

$$\frac{1}{q}\left(\frac{r^2}{q}\right)' + \frac{II'}{B_0^2} + \frac{4\pi R_0^2}{B_0^2} p' = 0. \tag{4.20}$$

where the safety factor q was introduced from $\psi' = rB_T/q \cong rB_0/q$.

Thus, only two of the three flux quantities q, I, and p may be specified independently. Next, we combine the terms $\propto \cos\vartheta$ in (4.17) and (4.18) to obtain:

$$\frac{1}{r}\left(2r\Delta'\psi'\right)' - \frac{1}{r}\psi'\left(\frac{r}{R_0} + \left(r\Delta'\right)'\right) + \frac{2r}{R_0}\frac{4\pi R_0^2}{\psi'} p' = 0, \tag{4.21}$$

which could be represented as

$$\left(r\Delta'\psi'^2\right)' = \frac{r}{R_0}\left(\psi'^2 - 8\pi r R_0^2{}' p'\right) \tag{4.22}$$

Solving this equation for the Shafranov shift derivative, we find

$$\Delta' = \frac{1}{r^2\psi'^2}\cdot\frac{r}{R_0}\int_0^r r\,\mathrm{d}r\left(\psi'^2 - 8\pi r R_0^2 p'\right) = \frac{r}{R_0}\left(\frac{\langle\psi'^2\rangle r}{2\psi'^2} - \frac{8\pi R_0^2}{\psi'^2}\left(p - \langle p\rangle\right)\right) = \frac{r}{R_0}\left(\frac{\langle B_p^2\rangle}{2B_p^2} - \frac{8\pi}{B_p^2}\left(\langle p\rangle - p\right)\right)$$

where

$$\langle p\rangle \equiv \frac{2}{r^2}\int_0^r r\,\mathrm{d}r\cdot p(r)$$

By introducing internal inductance,

$$l_i = \langle B_p^2\rangle / B_p^2, \tag{4.23}$$

and poloidal beta

$$\beta_p = \frac{8\pi}{B_p^2}\left(\langle p\rangle - p\right) \tag{4.24}$$

we obtain:

$$\Delta' = \frac{r}{R_0}\left(\frac{l_i}{2} + \beta_p\right). \tag{4.25}$$

4.3 SOME EXTRAORDINARY TOKAMAK EQUILIBRIA

Plasma equilibria in tokamaks are usually described by the Grad–Shafranov equation pretty well, so this approach is used in a majority of tokamak discharges in present-day machines. There are, however, some tokamak plasma scenarios, parameters of which correspond to pretty unusual limiting cases of the Grad–Shafranov approach. We consider two interesting scenarios here.

4.3.1 ADVANCED TOKAMAK SCENARIO WITH HOLLOW CURRENT PROFILES

The first extraordinary equilibrium to be considered is the so-called "advanced tokamak" (AT) scenarios aiming at obtaining internal transport barriers (ITBs) in tokamak plasmas [4.5–4.7]. The ITB scenarios are obtained by applying high power of an auxiliary heating to tokamak plasma early in the discharge and well before the inductive current diffuses to the plasma centre. Under certain conditions, a region of thermal plasma with much improved confinement properties rises at the plasma centre, the ITB. The very high ion temperature and plasma density inside the ITB delivers a very significant fusion performance of the plasma, for example, the record high rate of D-D neutrons, 5.5×10^{16}/s, was achieved on JET with carbon wall in the ITB scenario. Due to the incomplete penetration of the inductive current to the central region at the start of main heating, a hollow current profile is often created with non-monotonic $q(r)$-profile. In some cases, the plasma forms a "current hole" profile, with *zero* current value (within the error bars of measurements) over an extended central region of the plasma [4.8]. Figures 4.2–4.5 show an example of AT scenario in JET.

The temporal evolution of the D-D neutrons rate in Figure 4.2 (top) increases faster at $t=6$ s indicating formation of ITB, and the neutron rate doubles by 6.8 s at constant power of NBI and ICRH. Figure 4.3 shows how the magnetic topology changes via the safety factor $q(R)$ when the ITB forms, and Figures 4.4 and 4.5 show the profiles of ion temperature and electron density at the time slice corresponding to Figure 4.3. The flat profiles of plasma in the central region of the plasma are consistent with Eq. (4.14), which requires low values of plasma pressure gradient when the plasma current is small (safety factor is high) as in our case.

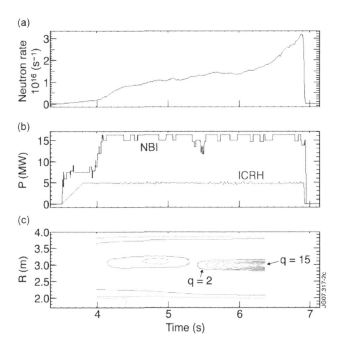

FIGURE 4.2 Temporal evolution of key discharge parameters in AT discharge JET #58094. From top to bottom: (a) Fusion yield measured via D-D neutrons; (b) NBI and ICRH power wave forms, and (c) isolines showing non-monotonic $q(r, t)$ profile.

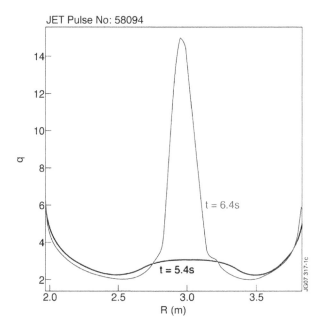

FIGURE 4.3 Profiles of the safety factor $q(R)$ measured in discharge #58094 at two different times. Equilibrium with a deeply reversed magnetic shear is formed prior to $t=6.4\,$s.

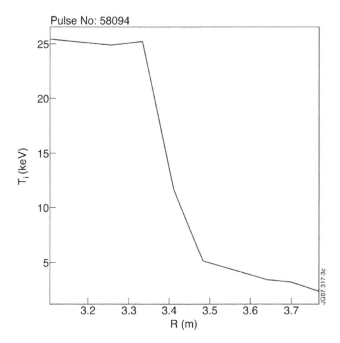

FIGURE 4.4 Profile of ion temperature $T_i(R)$ measured in the discharge #58094 at $t=6.4\,$s.

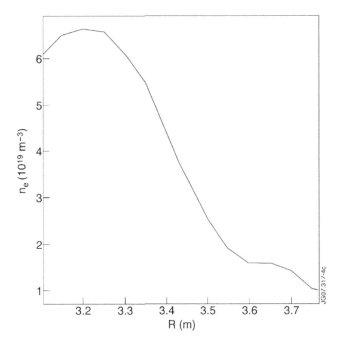

FIGURE 4.5   Profile of electron density $n_e(R)$ measured in the discharge #58094 at $t=6.4$ s.

In the case of a "current hole" equilibrium, central region of plasma becomes similar to a stellarator-type equilibrium without current surrounded by a tokamak-type equilibrium with finite current that twists (rotational transform) magnetic field lines. In this case, orbits of charged particles in the current hole region are determined by ∇B-drifts that cause the charge separation discussed in Chapter 3 followed by the $\boldsymbol{E} \times \boldsymbol{B}$-drift pushing plasma out of the higher magnetic field. However, a finite current density outside the current hole area causes a rotational transform that shortcut the separated charges and prevents the electric drift of the whole plasma.

4.3.2   Spherical Tokamaks with High-β

The second extraordinary equilibrium is the equilibrium obtained in "spherical tokamaks" (STs) at record high values of plasma β, $\beta \sim 1$. In this case, plasma pressure expels magnetic field from the centre of plasma and may form a diamagnetic well. The value of the Shafranov shift becomes comparable to the minor radius of tokamak, and the whole problem of solving the Grad–Shafranov equation could be separated [4.9] into a boundary value problem for plasma at the low field side, and a simplified one-dimensional description of the magnetic flux surfaces – at the high field side and the plasma center. Figure 4.6 shows an example of ST equlibrium with β (0) ~1 Figure 4.7 shows that a magnetic well is formed in this case.

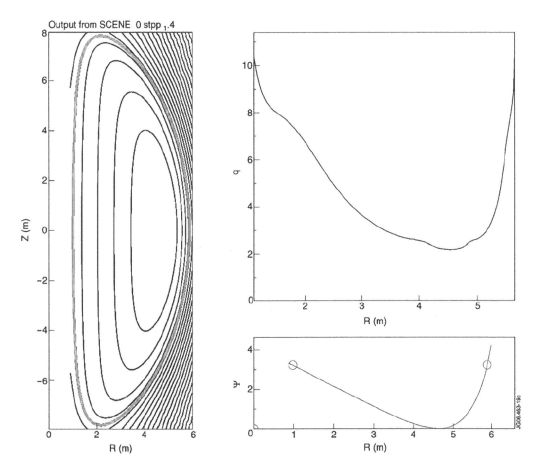

FIGURE 4.6   One-dimensional nature of the flux surfaces in the core of the plasma, together with the safety factor q and the flux function Ψ as functions of major radius R computed for STPP project [4.10].

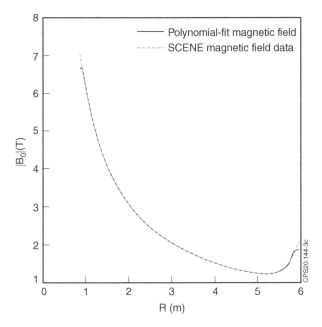

FIGURE 4.7   Radial profile of the total equilibrium magnetic field for the case in Figure 4.6.

REFERENCES

1. J.P. Freidberg, *Ideal magnetohydrodynamics*, Plenum Press, New York (1971).
2. V.S. Mukhovatov and V.D. Shafranov, *Nucl. Fusion* **11** (1971) 605.
3. V.D. Shafranov, *Soviet Phys. JETP* **6** (1958) 545.
4. H. Grad and H. Rubin, *Hydromagnetic Equilibria and Force-Free Fields*, Proceedings of the 2nd UN International Conf. on the Peaceful Uses of Atomic Energy, Geneva 1958 (Columbia University Press, New York, Vol. 31, 1959), 190.
5. T.S. Taylor, *Plasma Phys. Control. Fusion* **39** (1997) B47.
6. C.D. Challis et al., *Plasma Phys. Control. Fusion* **43** (2001) 861.
7. R.C. Wolf, *Plasma Phys. Control. Fusion* **45** (2003) R1.
8. N.C. Hawkes et al., *Phys. Rev. Lett.* **87** (2001) 115001.
9. S.C. Cowley et al., *Phys. Fluids* **B3**, (1991) 2066.
10. H.R. Wilson et al., *The Spherical Tokamak Fusion Power Plant*, 19th IAEA Fusion Energy Conference (IAEA, Lyon, France, 2002), No. FT/1-4.

5 MHD Waves in Magnetically Confined Plasmas

5.1 THE LINEARIZATION PROCEDURE AND MAIN TYPES OF MHD WAVES

We start from Eqs. (4.1)–(4.7) and use (4.13) for dividing plasma variables into equilibrium and perturbed quantities (denoted by δ) where all the perturbed quantities satisfy $\delta \ll 1$, that is, $\left|\dfrac{\delta J}{J_0}\right| \ll 1$, etc. The equilibrium terms are balanced owing to the plasma equilibrium, and now we consider only the terms linear in δ. The linearized ideal MHD equations take the form:

$$\frac{\partial \delta \rho}{\partial t} + \nabla \cdot \left(\rho_0 \delta V\right) = 0; \tag{5.1}$$

$$\rho_0 \frac{d \delta V}{dt} = -\nabla \delta p + \frac{1}{4\pi}\left[\nabla \times \delta B\right] \times B_0; \tag{5.2}$$

$$\frac{\partial}{\partial t}\delta B = \nabla \times \left[\delta V \times B_0\right]; \tag{5.3}$$

$$\delta p = \gamma \frac{p_0}{\rho_0}\delta \rho. \tag{5.4}$$

For clarifying the physics of the perturbations, we introduce a vector of plasma displacement from the equilibrium, ξ, related to δV via

$$\delta V = \frac{\partial \xi}{\partial t}, \tag{5.5}$$

so (5.1) and (5.3) take the form

$$\delta \rho = -\mathrm{div}\left(\rho_0 \xi\right); \tag{5.6}$$

$$\delta B = \nabla \times \left[\xi \times B_0\right] = -B_0 \mathrm{div}\,\xi_\perp + B_0 \frac{\partial \xi_\perp}{\partial z}, \tag{5.7}$$

where we used the vector analysis equation

$$\nabla \times \left[\mathbf{a} \times \mathbf{b}\right] = \left(\mathbf{b}\nabla\right)\mathbf{a} - \left(\mathbf{a}\nabla\right)\mathbf{b} + \mathbf{a}\,\mathrm{div}\,\mathbf{b} - \mathbf{b}\,\mathrm{div}\,\mathbf{a}, \tag{5.8}$$

and considered for simplicity a slab geometry case with

$$\boldsymbol{B}_0 \uparrow\uparrow \mathbf{e}_z. \tag{5.9}$$

By substituting these expressions for $\delta\rho, \delta\boldsymbol{B}$ in the remaining Eqs. (5.2) and (5.4), we obtain

$$\frac{\partial^2 \boldsymbol{\xi}}{\partial t^2} = c_S^2 \nabla \mathrm{div}\boldsymbol{\xi} + V_A^2 \nabla_\perp \mathrm{div}\boldsymbol{\xi}_\perp + V_A^2 \frac{\partial^2 \boldsymbol{\xi}_\perp}{\partial z^2}, \tag{5.10}$$

where

$$c_S^2 = \gamma \, p_0/\rho_0 \tag{5.11}$$

is square of ion sound speed, and

$$V_A^2 = B_0^2 / (4\pi\rho_0) \tag{5.12}$$

is squared Alfvén velocity.

Equation (5.10) describes linear MHD perturbations of homogeneous ideal conducting plasma. This equation for the single vector variable $\boldsymbol{\xi}$ gives three scalar equations for three different types of MHD waves [5.1]. Consider the first two "compressible" types of waves, in which $\xi_z \neq 0$ and $\mathrm{div}\,\boldsymbol{\xi}_\perp \neq 0$. The displacement ξ_z parallel to the equilibrium magnetic field is described by the parallel projection of (5.10):

$$\frac{\partial^2 \xi_z}{\partial t^2} = c_S^2 \frac{\partial^2 \xi_z}{\partial z^2} + c_S^2 \frac{\partial}{\partial z} \mathrm{div}\boldsymbol{\xi}_\perp, \tag{5.13}$$

and equation for $\mathrm{div}\boldsymbol{\xi}_\perp$ is obtained from the divergence of the perpendicular projection of (5.10):

$$\frac{\partial^2 \mathrm{div}\boldsymbol{\xi}_\perp}{\partial t^2} = c_S^2 \Delta_\perp \mathrm{div}\boldsymbol{\xi}_\perp + V_A^2 \left(\Delta_\perp + \frac{\partial^2}{\partial z^2} \right) \mathrm{div}\boldsymbol{\xi}_\perp + c_S^2 \Delta_\perp \frac{\partial \xi_z}{\partial z}, \tag{5.14}$$

where $\Delta_\perp = \mathrm{div}\nabla_\perp$. We see that (5.13) and (5.14) for ξ_z and $\mathrm{div}\boldsymbol{\xi}_\perp$ are coupled. However, if we consider the limit $\dfrac{c_S^2}{V_A^2} \ll 1$, Eq. (5.14) decouples from ξ_z and reduces to

$$\frac{\partial^2 \mathrm{div}\, \boldsymbol{\xi}_\perp}{\partial t^2} = V_A^2 \Delta_\perp \mathrm{div}\, \boldsymbol{\xi}_\perp. \tag{5.15}$$

Equation (5.15) describes compressional Alfvén (CA) wave, for which the magnetic pressure $B_0^2/8\pi$ determines the "returning" force that acts perpendicular to $\boldsymbol{B}_0$. The displacement ξ_z parallel to $\boldsymbol{B}_0$ is described by

$$\frac{\partial^2 \xi_z}{\partial t^2} = c_S^2 \frac{\partial^2 \xi_z}{\partial z^2}, \tag{5.16}$$

and the wave equation (5.16) describes ion sound wave existing in plasma even without $\boldsymbol{B}_0$.

However, if c_S^2/V_A^2 is not small, the two different compressible waves couple. In this case, the ion sound wave is modified by the magnetic pressure and becomes slow magneto-acoustic (SM) wave. The coupled equations give for $\xi_z, \mathrm{div}\boldsymbol{\xi}_\perp \propto \exp(-i\omega t + i\boldsymbol{k}\cdot\boldsymbol{r})$ the following dispersion relation between wave frequencies and wave vectors:

$$\omega^4 - \left(c_S^2 + V_A^2 \right) k^2 \omega^2 + V_A^2 c_S^2 k_z^2 k^2 = 0. \tag{5.17}$$

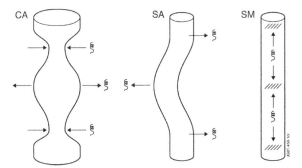

FIGURE 5.1 Plasma displacement ξ in compressional Alfvén (CA), shear Alfvén (SA), and slow magneto-acoustic (SM) wave.

In contrast to the compressional waves, the third type of MHD waves, the so-called shear Alfvén (SA) wave, is incompressible, $\xi_z = 0$ and $\mathrm{div}\,\xi_\perp = 0$. For such waves, the main MHD equation (5.10) simply becomes

$$\frac{\partial^2 \xi_\perp}{\partial t^2} = V_A^2 \frac{\partial^2 \xi_\perp}{\partial z^2},$$

(5.18)

which coincides with the well-known equation for oscillations of a string. The "returning" force for SA waves is the tension of magnetic field lines, which act similar to the strings.

The three different types of MHD waves derived above are schematically shown in Figure 5.1.

Figure 5.1 shows that CA and SM waves have "returning" forces as the magnetic and the kinetic pressure, respectively, while SA wave has the "returning" force as the tension of magnetic field lines.

5.2 SHEAR ALFVÉN WAVES IN INHOMOGENEOUS PLASMA AND THE WAVE CONTINUUM DAMPING

Among all MHD waves in plasma, SA wave constitutes the most significant part of the MHD spectrum and is probably the best studied [5.2]. In SA wave the fluid displacement vector ξ and perturbed electric field δE are perpendicular to the magnetic field B_0, while the wave propagates along B_0 with frequency and parallel wave-vector determined by

$$\omega = \pm k_\parallel V_A;$$

(5.19)

$$k_\parallel = \frac{k \cdot B_0}{B_0}.$$

(5.20)

However, wave packets with dispersion relation (5.19) are not structurally stable in inhomogeneous plasmas. Let us consider plasma with inhomogeneous density, $V_A = V_A(r)$, and finite magnetic shear, that causes a radial dependence in parallel wave-vector, $k_\parallel = k_\parallel(r)$, and assume that a wave-packet of SA type is created by some means at $t = t_0$ as Figure 5.2 shows. If the wave-packet satisfies SA dispersion relation (5.19) in each point along the axis r, and both Alfvén velocity and parallel wave-vector do vary along r too, then every individual "slice" of the wave-packet at different r will propagate with a different phase velocity (due to the r-dependence in Alfvén velocity) and in a different direction (due to the r-dependence in the parallel wave-vector).

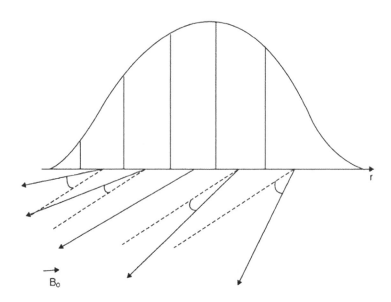

FIGURE 5.2   A wave-packet of SA type extended over r-variable in inhomogeneous plasma. The arrows show the directions of $\boldsymbol{B}_0$, which change along r due to the assumed magnetic shear.

After some time, the slices of the wave-packet with have their phases shifted significantly enough for spreading the wave-packet along magnetic field. The life-time of the wave-packet of SA waves is limited by the "phase mixing"

$$\tau^{-1} \propto \frac{\mathrm{d}}{\mathrm{d}r}\left(k_\parallel(r)\cdot V_A(r)\right). \tag{5.21}$$

In addition to the initial value problem above, it is instructive to consider a boundary value problem with SA wave launched by an external antenna into an inhomogeneous cold plasma with density gradient, $n_0 = n_0(x)$, $P_0 = 0$, $\boldsymbol{B}_0 = B_0 \mathbf{e}_Z$, as shown in Figure 5.3.

An externally excited electromagnetic wave with perturbed electrostatic potential $\phi \propto \phi(x)\exp(ik_y y - i\omega t)$ is described in our case by equation

$$\frac{\mathrm{d}}{\mathrm{d}x}\left(\omega^2 - \omega_A^2(x)\right)\frac{\mathrm{d}\varphi}{\mathrm{d}x} - k_y^2\left(\omega^2 - \omega_A^2(x)\right)\varphi = 0, \tag{5.22}$$

where

$$\omega_A^2(x) \equiv k_\parallel^2 V_A^2(x). \tag{5.23}$$

Equation (5.22) has a zero bracket at the high-order derivative term at the point $x = x_0$ of the local Alfvén resonance layer, where frequency of the externally launched wave matches the local SA frequency,

$$\omega^2 = \omega_A^2(x_0). \tag{5.24}$$

The equation in the vicinity of this point can be simplified to:

$$\frac{\mathrm{d}}{\mathrm{d}x}\left(\omega^2 - \omega_A^2(x)\right)\frac{\mathrm{d}\varphi}{\mathrm{d}x} = 0, \tag{5.25}$$

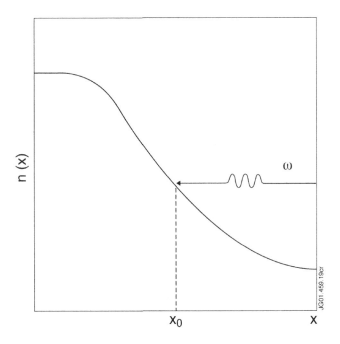

FIGURE 5.3 Wave is launched with frequency ω and wave vector $k_{\parallel}$ in a slab geometry from low-density plasma edge into inhomogeneous plasma with density gradient.

which after integration gives

$$\frac{d\varphi}{dx} = \frac{\text{const}}{\omega^2 - \omega_A^2(x)}. \tag{5.26}$$

and expansion in the vicinity of the point $x = x_0$:

$$\omega^2 \approx \omega_A^2(x_0) + \left(d\omega_A^2(x)/dx\right)_{x=x_0} \cdot (x - x_0) \tag{5.27}$$

gives the following solutions:

$$\varphi \propto \text{const} \cdot \ln(x - x_0),\ x > x_0 \tag{5.28}$$

$$\varphi \propto \text{const} \cdot \left(\ln|x - x_0| + i\pi\right),\ x < x_0 \tag{5.29}$$

The singularity at $x = x_0$ shows that the energy of the launched wave peaks at $x = x_0$, while the term $i\pi$ shows that resonant absorption of the wave energy occurs at this point, the continuum damping of the wave launched.

Due to the large values of the continuum damping, excitation of SA wave by super-Alfvénic energetic ions was not considered for a while as a major threat to the burning plasma operation. However, this situation changed in the 1980s when global Alfvén eigenmodes (GAEs) and toroidal Alfvén eigenmodes (TAEs) were reported.

5.3 DISCOVERY OF GLOBAL ALFVÉN EIGENMODES IN CYLINDRICAL PLASMAS

Several research groups investigated possible heating of plasma with externally launched waves of the SA frequency range. It was found in numerical modelling that in cylindrical plasmas, as shown in Figure 5.4, in addition to the continuous Alfvén spectrum,

$$\omega^2 = \omega_A^2(r) \equiv k_\parallel^2(r) V_A^2(r), \tag{5.30}$$

a discrete GAE with frequency $\omega_{GAE} < \omega_A$ exists if plasma has a current parallel to the magnetic field [5.3,5.4]. Note that the cylindrical geometry naturally implies a finite (non-zero) parallel wave vector determined by the cylinder's length L, $k_\parallel^{min} = 2\pi L^{-1} > 0$. This also determines the lowest frequency of the SA continuum in cylindrical geometry, $\omega_A^2 \equiv \left(k_\parallel^{min}\right)^2 V_A^2 > 0$, below which SA waves determined by Eq. (5.12) cannot exist. This is the frequency range where GAE exists without experiencing coupling to the waves of the SA continuum.

A new high-quality, $Q \equiv \omega/\gamma \sim 10^3$, resonance was discovered during these Alfvén antenna studies in plasmas with current, as shown in Figure 5.5.

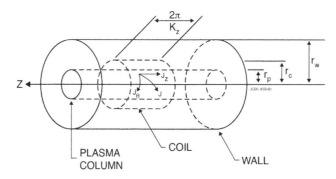

FIGURE 5.4   Geometry of the plasma, coil, and the wall cylindrical system used in the numerical investigation[*].

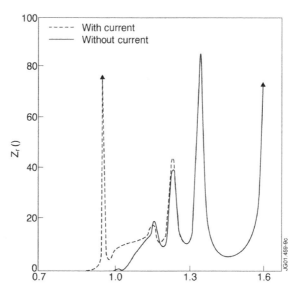

FIGURE 5.5   Real part of the computed coil impedance versus normalized frequency $\omega/(k_\parallel^{min} \cdot V_A)$. A new resonance with a very high quality is detected just below the continuum frequency in plasma with current[*].

---

[*] Reproduced from [D.W. Ross et al. *Phys. Fluids* 25 (1982) 652], with the permission of AIP Publishing.

The theoretical interpretation of the new discrete frequency GAE was provided. It was shown that GAE exists in ideal MHD if the current profile determining dB_ϑ/dr provides a minimum in Alfvén continuum, that is,

$$\left(d\omega_A\left(r\right)/dr\right)_{r=r_0}=0 \tag{5.31}$$

which can be expressed as a condition for parallel wave vector at a given plasma density profile:

$$\frac{1}{k_\parallel}\frac{dk_\parallel}{dr}=-\frac{1}{V_A}\frac{dV_A}{dr}. \tag{5.32}$$

Figure 5.6 shows that all continuum curves for various poloidal mode numbers had minima in the modelling that included plasma current.

The local minimum of Alfvén continuum provides a maximum of the perpendicular refraction index $N_r = ck_r/\omega$ that describes electromagnetic waves (and SA wave is an electromagnetic wave). In the presence of the maxima in the refraction index, similar to the fibre optics, the wave tends to propagate in a "wave-guide" surrounding the region of the extremum refraction index. Figure 5.7 shows that the structure of GAE peaks in the vicinity of the extremum point satisfying (5.31). Furthermore, we see no logarithmic singularity of the type (5.28), (5.29) in the computed GAE. This means that although GAE is a SA perturbation, it does not experience the continuum damping. The most significant damping typical of the SA wave packets satisfying (5.19), does not exist for GAE as its eigenfrequency does not cross the local Alfvén frequency anywhere in the plasma. Because GAE frequency does not satisfy the local Alfvén resonance condition,

$$\omega_{\mathrm{GAE}} \neq \omega_A\left(r\right),\quad 0<r/a<1, \tag{5.33}$$

this SA mode has no singularity, does not experience the continuum damping, and belongs to weakly damped Alfvén Eigenmodes (AEs) with high quality factors.

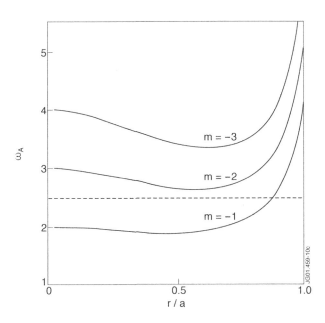

FIGURE 5.6 Radial structure of Alfvén continuum in cylindrical plasmas with current. Poloidal mode numbers from −1 to −3 have minimum points.

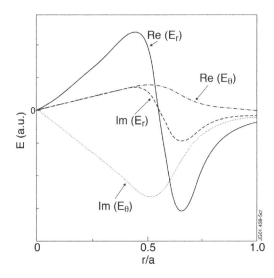

FIGURE 5.7   Radial structure of the computed ideal MHD GAE with $m=-2$.

5.4   KINETIC ALFVÉN WAVES VERSUS SHEAR ALFVÉN WAVES

Kinetic Alfvén waves (KAWs) [5.5] represent the most important branch of waves that can couple with SA waves in inhomogeneous plasmas, and which often determine the microscopic mechanisms of the damping of SA waves. KAWs are associated with finite Larmor radius effects, which play an important role when the perpendicular wavelength, $2\pi/k_\perp$, becomes comparable to the Larmor radius of thermal ions. In such short wavelength perturbations, ions can no longer follow the magnetic field lines, whereas electrons are still frozen in the field lines because of their small Larmor radius. This produces charge separation and additional parallel electric field.

In particular, the continuum damping of SA wave, which results from the singularity in the wave amplitude at the resonance position (5.23) in ideal MHD in real plasma corresponds to the wave transformation into KAW. Figure 5.8 shows a comparison of the mode structures near Alfvén resonance position for the ideal MHD approximation and KAW. In this figure, an SA wave launched from the left side approaches the Alfvén resonance position at $X=0$. As it does so, the wave amplitude increases and gives in ideal MHD model a singularity and the wave continuum damping in accordance with (5.28) and (5.29). In real plasma, KAWs are excited as soon as the distance between the wave launched and the resonance point becomes comparable to ion Larmor radius. The wavelength of KAW is of the order of the Larmor radius, and the wave has an oscillatory structure propagating across the magnetic field.

The dispersion relation of KAW has the form

$$\omega^2 = k_\parallel^2 V_A^2 \left(1 + k_\perp^2 \rho_i^2 \left(\frac{3}{4} + \frac{T_e}{T_i} \right) \right). \tag{5.34}$$

In contrast to SA wave, KAW can propagate across the magnetic field, and has electric field components not only across the magnetic field but also in the direction of the magnetic field. Note also that this wave propagating through a region with $k_\parallel = 0$ (e.g., across rational magnetic surfaces in tokamaks) becomes of an infinitely short wavelength, $k_\perp \to \infty$, since its frequency is constant. This makes KAW modelling difficult in real space, unless an absorption of some kind increasing with $k_\perp$ exists for the wave, which can cut off the very short wavelengths.

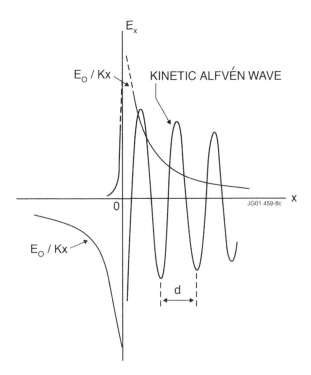

FIGURE 5.8   Structure of kinetic Alfvén wave in the vicinity of the Alfvén resonance.

Finally, we note that in plasmas with ordering $V_{Te} \ll V_A$, which is rare in tokamaks, the electron inertia becomes important and the KAW dispersion relation is modified to

$$\omega^2 = k_{\parallel}^2 V_A^2 \frac{\left(1 + k_{\perp}^2 \rho_i^2\right)}{\left(1 + \dfrac{k_{\perp}^2 c^2}{\omega_{pe}^2}\right)}. \tag{5.35}$$

REFERENCES

1. B.B. Kadomtsev, *Kollektivnye yavleniia v plazme*, NAUKA, Moscow (1976) (in Russian).
2. A. Hasegawa and C. Uberoi, *The Alfven Wave*. A.Hasegawa and C.Uberoi. Technical Information Center, U.S. Department of Energy, (1982).
3. D.W. Ross et al. *Phys. Fluids* **25** (1982) 652.
4. K. Appert et al. *Plasma Phys.* **24** (1982) 1147.
5. A. Hasegawa and L. Chen, *Phys. Rev. Lett.* **32** (1974) 454.

6 Toroidicity-Induced Alfvén Eigenmodes (TAEs)

In the mid-1980s, magnetic fusion research reached a point at which large-scale thermonuclear plasmas with significant populations of energetic ions became possible. Several large tokamaks were built world-wide, and powerful auxiliary heating systems, neutral beam injection (NBI) and ion cyclotron resonance heating (ICRH), generating energetic ions were developed successfully. Two of the large tokamaks, TFTR (United States) and JET (United Kingdom), were designed to operate with DT plasmas, and hence, experimental studies of fusion-generated alpha particles in tokamak plasmas became possible.

Therefore, it is essential to consider in detail what novel physics features may arise from the presence of large populations of energetic ions in tokamak plasmas, especially if these populations were super-Alfvénic (i.e. speeds of the energetic ions exceeded the Alfvén speed). Theoretically, it has been pointed out much earlier in Refs. [6.1,6.2] that coupling of super-Alfvénic ions to the Alfvén waves might excite Alfvén instabilities. Such instabilities, if the wave amplitudes would become sufficiently high, could result in a cross-field transport of the energetic ions much higher than the neoclassical transport due to Coulomb collisions. Consequently, due to the Alfvén instabilities, plasma heating by the energetic ions could decrease significantly and change the power deposition profile. Moreover, enhanced losses of the energetic ions, if repeated discharge after discharge, could damage the first wall.

However, a principal uncertainty in calculating the critical threshold in energetic particle pressure for exciting Alfvén instabilities was the strong damping of shear Alfvén waves. As we saw in Chapter 5, in a sheared magnetic field in a slab geometry, these waves are highly localised at the surfaces $\omega = k_\parallel V_A$ and experience strong continuum damping. An exception was found for global Alfvén eigenmodes (GAEs), but such modes existing in cylindrical geometry with plasma current still experience a strong continuum damping in toroidal geometry [6.3].

Therefore, a dedicated search for weakly damped Alfvén modes in toroidal geometry was performed. This resulted in the discovery of toroidal Alfvén eigenmodes (TAE) [6.4]. TAE exists due to the periodic nature of the toroidal field, which gives rise to gaps in the Alfvén continuum frequency spectrum, within which a discrete spectrum of TAEs may result from a finite magnetic shear. TAEs are discrete modes free of the continuum damping in the lowest order, and can be easily destabilised by alpha particles or other energetic ions [6.5,6.6].

The predicted weakly damped TAE modes residing inside the continuum gaps with a well-determined frequency were searched for in tokamak experiments. The instability of TAE was observed for the first time in TFTR and DIII-D experiments with NBI-produced energetic ions injected into plasmas with low magnetic field, $B \leq 1$ T, so that V_A was comparable to the velocities of the beam ions [6.7,6.8]. Significant loss of the beam ions, up to 45% on DIII-D [6.8], was observed, indicating how crucial TAE instabilities could be in the presence of large population of energetic ions with velocities close to V_A. These early experiments made the TAE issue one of the highest priorities for magnetic fusion studies world-wide. In addition, it was found in Refs. [6.7,6.8] that the experimentally observed thresholds for TAE instabilities were higher than those predicted theoretically. This attracted much attention to the theoretical studies of TAE damping and drive, but with more precision.

In present-day experiments, TAEs excited by several types of energetic ion populations are often observed in many tokamaks with auxiliary heating. In particular, TAEs driven by fusion-generated alpha-particles were detected in TFTR DT experiment [6.9]. Furthermore, several new types of weakly damped Alfvén eigenmodes (AEs) were found, including the "gap" modes with frequencies within the gaps in the continua created by the ellipticity and triangularity of the plasma cross-section, and/or the geodesic curvature and plasma compressibility. Other types of present-day magnetic fusion machines, such as stellarators and reversed-field pinches, also often observe Alfvén instabilities in discharges with auxiliary heating.

In this chapter and in Appendix C we present an analytical theory of TAE in the large aspect ratio tokamak with low magnetic shear, as well as qualitatively describe the main effects that determine TAE drive by energetic particles and TAE damping due to thermal plasma. We believe this would facilitate studies of other types of weakly damped AEs by using the TAE description here as an example.

6.1   ANALYTIC THEORY OF TAE

The ideal MHD approach gives the following equation for shear Alfvén waves describing a balance between *the bending energy term* (first term in the equation) and *the inertial term* (second term in the equation):

$$
\boldsymbol{B} \cdot \nabla\left(\frac{1}{B} \nabla_\perp^2 (\boldsymbol{b} \cdot \nabla \phi) \right) + \nabla \cdot \left(\frac{\omega^2}{v_A^2} \nabla_\perp \phi \right) = 0 \tag{6.1}
$$

Here, the plasma pressure gradient is assumed to be negligibly small, $-\dfrac{8\pi R q^2}{B^2} \cdot \dfrac{\mathrm{d}p}{\mathrm{d}r} \ll S^2$. The derivation of (6.1) is given in Appendix C. To analytically investigate this three-dimensional equation, we transform it to a set of one-dimensional coupled equations by following the approach developed in Ref. [6.10]. This is done by taking into account that the wave solutions in bounded toroidal plasmas should be periodic and quantised in toroidal and poloidal directions:

$$
\phi\left(r,\ \vartheta,\ \zeta,\ t\right) = \exp(-i\omega t + in\zeta) \sum_m \varphi_m(r) \exp(-im\vartheta) + c.c., \tag{6.2}
$$

where n is the number of wavelengths in toroidal direction, m is the number of wavelengths in poloidal direction, and *c.c.* stands for "complex conjugate."

By assuming concentric magnetic flux surfaces, low magnetic shear, and taking only the first-order inverse aspect ratio expansion, $\dfrac{r}{R} \ll 1$, we retain two dominant poloidal harmonics for the mode, m-the and $(m-1)$-th, and obtain (see Appendix C) two coupled second-order differential equations for the amplitudes $\varphi_m(r)$, $\varphi_{m-1}(r)$ of the wave electrostatic potential:

$$
L_m \varphi_m + \varepsilon \frac{1}{4q^2 R^2} \frac{\mathrm{d}^2}{\mathrm{d}r^2} \varphi_{m-1} = 0 \tag{6.3}
$$

$$
L_{m-1} \varphi_{m-1} + \varepsilon \frac{1}{4q^2 R^2} \frac{\mathrm{d}^2}{\mathrm{d}r^2} \varphi_m = 0, \tag{6.4}
$$

where the toroidicity coupling coefficient is $\varepsilon = \left(\dfrac{5}{2} \right)\left(\dfrac{r}{R} \right)$, and the differential operator L_m is defined as

$$
L_m \varphi_m \equiv \frac{\mathrm{d}}{\mathrm{d}r}\left(\frac{\omega^2}{V_A^2} - k_{\parallel m}^2 \right)\frac{\mathrm{d}\varphi_m}{\mathrm{d}r} - \frac{m^2}{r^2}\left(\frac{\omega^2}{V_A^2} - k_{\parallel m}^2 \right)\varphi_m. \tag{6.5}
$$

The set of Eqs. (6.3)–(6.5) describes TAEs.

In toroidal geometry with plasma current, the parallel wave-vector of the m-th harmonic of a wave with toroidal mode number n has the form

$$k_{\|m}(r) = \frac{1}{R}\left(n - \frac{m}{q(r)}\right), \qquad (6.6)$$

and its radial dependence is determined by the safety factor $q(r) = rB_\zeta / RB_\vartheta$. Two important properties of the parallel wave-vector are essential for further analyses. First, the parallel wave-vector (6.6) could be zero, in contrast to the cylindrical geometry. Indeed, if for some values of q (taken at a certain radius r), and integers n, m the condition $m = nq$ is valid, a rational magnetic surface with $k_m = 0$ exists at this radius. This implies that the frequency of the shear Alfvén continuum starts from zero, in contrast to the cylindrical plasma case. Second, two parallel wave-vectors of neighbouring poloidal mode numbers, m and $m-1$, could be equal at the same radial position satisfying

$$q = (m - 1/2)/n, \qquad (6.7)$$

thus making the continuum spectrum degenerate. Let us explain this specific property of toroidal geometry in detail.

In cylindrical geometry, $\frac{r}{R} \to 0$, two poloidal modes $\varphi_m(r)$ and $\varphi_{m-1}(r)$ are decoupled, as shown in Figure 6.1 (left) with two broken lines. Equations (6.3) and (6.4) become singular at $\omega_1^2 = k_{\|m}^2 V_A^2$ and $\omega_2^2 = k_{\|m-1}^2 V_A^2$, which give the two cylindrical shear Alfvén continua. In toroidal geometry, Eqs. (6.3) and (6.4) become coupled due to the finite toroidicity, as the poloidal mode numbers are not good quantum numbers any longer. The shear Alfvén continuum in toroidal geometry is obtained by setting the determinant of the coefficients in front of the second-order derivatives in (6.3), (6.4) equal to zero, that is,

$$\left(\frac{\omega^2}{V_A^2} - k_{\|m}^2\right)\cdot\left(\frac{\omega^2}{V_A^2} - k_{\|m-1}^2\right) - \frac{\varepsilon^2}{16q^4 R^4} = 0. \qquad (6.8)$$

This gives the following two branches:

$$\omega_\pm^2 = \frac{1}{2}\left(k_{\|m}^2 V_A^2 + k_{\|m-1}^2 V_A^2 \pm \left[\left(k_{\|m}^2 V_A^2 - k_{\|m-1}^2 V_A^2\right)^2 + \left(\frac{\varepsilon}{2}\right)^2\left(\frac{V_A}{qR}\right)^4\right]^{1/2}\right). \qquad (6.9)$$

At the point where the two cylindrical continua would cross,

$$k_{\|m-1}^2 V_A^2 = -k_{\|m}^2 V_A^2, \qquad (6.10)$$

a gap in the toroidal shear Alfvén continuum appears with the frequency width

$$\Delta\omega = \omega_+ - \omega_- \cong \frac{\varepsilon V_A}{(2qR)}. \qquad (6.11)$$

The toroidicity-induced gap in the frequency of shear Alfvén continuum has two extrema, at the bottom and the top of the gap, satisfying

$$\left.\frac{\mathrm{d}\omega_A(r)}{\mathrm{d}r}\right|_{r=r_0} = 0, \qquad (6.12)$$

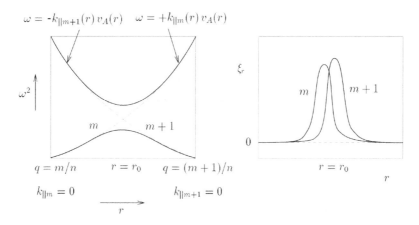

FIGURE 6.1 Left: Radial structure of two cylindrical Alfvén continua in toroidal plasma. Right: two poloidal harmonics form an eigenmode localised around radius satisfying Eq. (6.9), with the eigenmode frequency inside the TAE-gap.

similar to the case of GAE in cylindrical plasma with current. As in the case of GAE, we can expect that the toroidicity-induced extrema (6.12) and the corresponding extrema in the perpendicular refraction index could cause a wave-guide at the radius r_0, where an eigenmode may be formed.

To find discrete TAE spectrum from Eqs. (6.3) and (6.4), we introduce a normalised frequency as follows:

$$g = \frac{1}{\varepsilon}\left(\omega^2 \left(\frac{2qR}{V_A} \right)^2 - 1 \right), \tag{6.13}$$

so this frequency is $g = 0$ at the centre of the TAE-gap where the two cylindrical continua cross, (6.10), and $g = \pm 1$ at the bottom and the top of the TAE-gap. We will now look for a discrete TAE eigenfrequency inside the TAE-gap frequency. In the case of a low magnetic shear, such frequency is close to the bottom tip of the continuum [6.11]

$$g \approx -1 + \delta g, \quad 0 < \delta g \ll 1. \tag{6.14}$$

Figure 6.2 (right) explains the structure of the gap in the normalised frequency units.

Next, we will consider two radial regions of Eqs. (6.3) and (6.4): the inner region in the vicinity of the extremum point determined by (6.10), and the outer region away from (6.10). In the inner region, the effect of toroidicity is essential and the radial derivatives exceed significantly the poloidal scale of the wave variation, $\left| \dfrac{d}{dr} \right| \gg \dfrac{m}{r}$, so that last term in (6.5) could be neglected. Eqs. (6.3) and (6.4) can be integrated in the inner region to give

$$\phi_m^{in} = \frac{C_m}{2} \ln\left| z^2 - \left(1 - g^2\right) \right| - \frac{gC_m + C_{m-1}}{\left(1 - g^2\right)^{1/2}} \tan^{-1}\left(\frac{z}{\left(1 - g^2\right)^{1/2}} \right) + \text{const}$$

$$\tag{6.15}$$

$$\phi_{m-1}^{in} = \frac{C_{m-1}}{2} \ln\left| z^2 + \left(1 - g^2\right) \right| + \frac{C_m + gC_{m-1}}{\left(1 - g^2\right)^{1/2}} \tan^{-1}\left(\frac{z}{\left(1 - g^2\right)^{1/2}} \right) + \text{const}$$

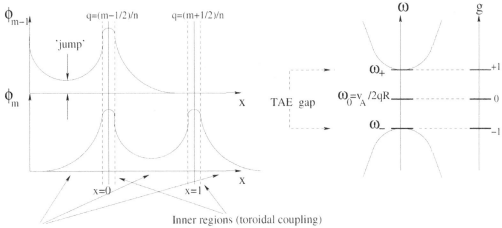

FIGURE 6.2 Left: Radial structure of two coupled poloidal harmonics of TAE. Right: the normalised frequency g in the vicinity of a TAE-gap.

where

$$z = \frac{4}{\varepsilon}\left(nq(r) - m + 1/2\right) \tag{6.16}$$

and C_m, C_{m-1} are the integration constants.

The symmetric (with respect to $z \to -z$ transform) logarithmic parts of the inner solutions (6.15) are the remnants of the logarithmic solutions that gave the singularity at the continuum resonance in cylindrical plasmas. Now, there is no logarithmic singularity in toroidal geometry because of the choice of the eigenfrequency (6.14) inside the gap in the continuum.

The asymmetric parts of the type $\tan^{-1}(a \cdot z)$ of the inner solutions are caused by the involvement of every poloidal harmonic in the radial regions at to two neighbouring gaps. These asymmetric parts give jumps of the inner solutions:

$$\Delta \phi_m^{in} = \phi_m^{in}(z \to +\infty) - \phi_m^{in}(z \to -\infty) = -\frac{gC_m + C_{m-1}}{\left(1 - g^2\right)^{1/2}}\left(+\frac{\pi}{2} - \left(-\frac{\pi}{2}\right)\right) = -\pi \frac{gC_m + C_{m-1}}{\left(1 - g^2\right)^{1/2}} \tag{6.17}$$

$$\Delta \phi_{m-1}^{in} = \ldots = \pi \frac{C_m + gC_{m-1}}{\left(1 - g^2\right)^{1/2}}$$

These jumps should be compensated by the outer parts of the solutions as there must be no asymmetry in the TAE electrostatic potential at $z \to \pm\infty$.

The outer solution describes the "cylindrical" harmonics away from the region of toroidal coupling. The outer solution is obtained from Eqs. (6.3)–(6.5) by neglecting toroidity and taking $\left|\frac{d}{dr}\right| \sim \frac{m}{r}$, which gives

$$x\frac{d^2\varphi_m}{dx^2} + (1-x)\frac{d\varphi_m}{dx} - x\frac{\varphi_m}{S^2} = 0, \tag{6.18}$$

$$x\frac{d^2\varphi_{m-1}}{dx^2} + (1+x)\frac{d\varphi_{m-1}}{dx} - x\frac{\varphi_{m-1}}{S^2} = 0, \tag{6.19}$$

where $x = (\varepsilon/4)z$.

The outer localised solutions of Eqs. (6.18) and (6.19) are:

$$\varphi_m = D_m \exp\left(-\frac{|x|}{S}\right) U\left(\frac{1}{2} - \frac{S}{4}\mathrm{sgn}(x), 1, \frac{2|x|}{S}\right),$$ (6.20)

$$\varphi_{m-1} = D_{m-1} \exp\left(-\frac{|x|}{S}\right) U\left(\frac{1}{2} + \frac{S}{4}\mathrm{sgn}(x), 1, \frac{2|x|}{S}\right),$$ (6.21)

where D_m, D_{m-1} are constants, and $U(a, b, cx)$ is the confluent hypergeometric function. For matching the inner layer solutions, we take the asymptotics of (6.20), (6.21) for $|x|/S \ll 1$:

$$\varphi_m \to \pi^{-1/2} D_m \left(\ln\left(\frac{2|x|}{S}\right) + \gamma - 2\ln 2 - \frac{\pi^2 S}{8}\mathrm{sgn}(x)\right),$$ (6.22)

$$\varphi_{m-1} \to \pi^{-1/2} D_{m-1} \left(\ln\left(\frac{2|x|}{S}\right) + \gamma - 2\ln 2 + \frac{\pi^2 S}{8}\mathrm{sgn}(x)\right),$$ (6.23)

where γ is the Euler constant.

In the limit of low magnetic shear, $S \ll 1$, the confluent hypergeometric function can be expressed via the Bessel function:

$$U\left(\frac{1}{2}, 1, \frac{2|x|}{S}\right) = \pi^{-1/2} \exp\left(\frac{|x|}{S}\right) K_0\left(\frac{|x|}{S}\right).$$ (6.24)

The jumps in the inner layer, (6.17), compensated by the jumps in the outer region, (6.22) and (6.23), give the dispersion relation for TAE with the same sign of two poloidal harmonics, that is, $C_m \approx C_{m-1}$:

$$\delta g \cong \frac{\pi^2 S^2}{8},$$ (6.25)

which transforms to TAE frequency by taking into account Eqs.(6.13) and (6.14):

$$\omega_{\mathrm{TAE}} = \frac{V_A}{2qR}\left(1 - \frac{\varepsilon}{2}\left(1 - \frac{\pi^2 S^2}{8}\right)\right).$$ (6.26)

In plasmas with ellipticity and triangularity of the cross-sections, double and triple oscillations in the poloidal angle appear in the components of the metric tensor (these oscillations were not considered in Appendix B). The double and triple oscillations give rise to other types of the "gap" modes in the shear Alfvén continuum in addition to the toroidicity-induced gap. Namely, ellipticity-induced gaps and ellipticity-induced Alfvén eigenmodes (EAEs) exist due to the coupling between poloidal m-th and $(m+2)$-th harmonics, and triangularity-induced gaps and non-circular (triangular) Alfvén eigenmodes (NAEs) exist due to the coupling between poloidal m-th and $(m+3)$-th harmonics. Figure 6.3a shows an example of a computed spectrum of waves in a typical JET plasma with ellipticity and some triangularity, with compressibility effects taken into account. The Alfvén spectrum is a mixture of continuum bands and gaps in this continuum, inside which discrete eigenfrequencies exist corresponding to TAEs and EAEs. At low frequency, a computed ion sound continuum is seen. Figure 6.3b shows that computed radial structure of TAE in the plasma with elliptical cross-section adds more poloidal harmonics. An analytical theory of EAE in a low-shear tokamak

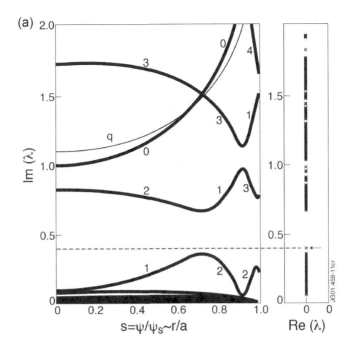

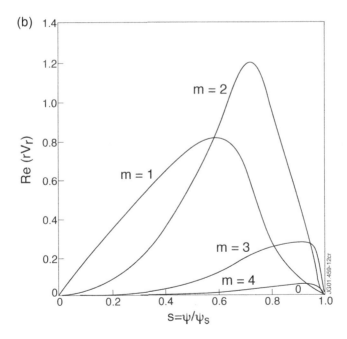

FIGURE 6.3 (a) Left: Radial structure of shear Alfvén continuum with $n = 1$, $Im(\lambda) = Im(\gamma R_0 / V_A(0))$, in a typical JET discharge. Right: the shear Alfvén spectrum corresponding to $Re(\lambda) = 0$ (waves, no instabilities). (b) Radial structure of $n = 1$ toroidal Alfvén eigenmode continuum with frequency marked by the arrow in (a). In addition to the two dominant poloidal harmonics of $m = 1$ and $m = 2$, higher poloidal harmonics are also coupled due to the large magnetic shear and non-circular plasma cross-section.

was developed similar to TAE, with the ordering $\left|\dfrac{\mathrm{d}}{\mathrm{d}r}\right| \gg \dfrac{m}{r}$ and two starting equations for the amplitudes $\varphi_m(r)$, $\varphi_{m-2}(r)$ of the electrostatic potential [6.12]:

$$L_m \varphi_m + \frac{e}{q^2 R^2}\frac{\mathrm{d}^2}{\mathrm{d}r^2}\varphi_{m-2} = 0, \tag{6.27}$$

$$L_{m-2} \varphi_{m-2} + \frac{e}{q^2 R^2}\frac{\mathrm{d}^2}{\mathrm{d}r^2}\varphi_m = 0, \tag{6.28}$$

where the ellipticity coupling coefficient is $e = \left(E^2 - 1\right)/\left(E^2 + 1\right)$, $E \equiv b/a$, and a, b are the lengths of the elliptical semi-axes.

6.2 KINETIC ALFVÉN WAVES IN THE VICINITY OF TAE-GAP: QUALITATIVE PICTURE OF RADIATIVE DAMPING OF TAE AND DISCRETE SPECTRUM OF KINETIC TAE

It was pointed out in Ref. [6.13] that TAE may have quite significant coupling to kinetic Alfvén waves (KAWs) as the TAE radial structure peaks at the position of the TAE-gap and the Larmor radius parameter, $k_\perp^2 \rho_i^2$, is not negligibly small. To explain the possible results of the coupling between TAE and KAW, we consider Figure 6.4 showing Alfvén continuum for two coupled poloidal harmonics, and the regions of KAW propagation obtained in the first-order Larmor radius approximation keeping the terms $\propto k_\perp^2 \rho_i^2$.

In toroidal geometry, KAWs of the same toroidal mode number n should be labelled in accordance with their poloidal mode numbers, so that a KAW with harmonic m is described by the dispersion relation:

$$\omega^2 = k_{\|m}^2 V_A^2 \left(1 + k_\perp^2 \rho_i^2 \left(\frac{3}{4} + \frac{T_e}{T_i}\left(1 - i\delta_e\right)\right)\right). \tag{6.29}$$

Here, $k_\perp^2 = k_r^2 + k_\vartheta^2$, $\rho_i^2 = \dfrac{2T_i}{m_i \omega_{Bi}^2}$ is the square of the thermal ion Larmor radius, and the small dissipative part, $\delta_e \ll 1$, is caused by trapped electron Coulomb collisions with passing electrons [6.14].

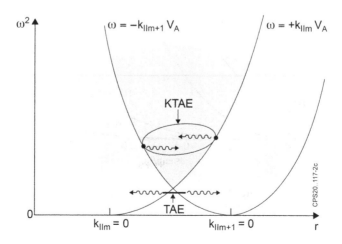

FIGURE 6.4 Radial structure of toroidal Alfvén continuum with poloidal harmonics m and $m+1$ (same n) and the areas of propagation of KAWs with harmonics m and $m+1$ (shaded areas).

In contrast to shear Alfvén waves that do not propagate across the magnetic field, $\partial\omega/\partial k_r = 0$, KAWs have $\partial\omega/\partial k_r \neq 0$, so that they propagate across the magnetic field. It is important to identify qualitatively, e.g. from Figure 6.4, which ways the cross-field KAW energy fluxes can go. The sign "+" in front of the finite Larmor radius term in (6.29) tells us that KAW with poloidal mode number m can only exist and propagate *above* the Alfvén continuum curve with the same m. These areas of KAW propagation are shown in grey in Figure 6.4. There are two different areas in the vicinity of a TAE-gap: one for KAW with mode number m, and the other – for KAW with mode number $m+1$.

A significant asymmetry in KAWs exists between the frequency regions above and below the central frequency of the TAE-gap, $\omega_0 = 1/(2qR)$. Below the centre ω_0, outgoing fluxes of KAWs with poloidal mode numbers m and $m+1$ propagate radially away from each other, as well as away from the centre of TAE-gap localisation. If a TAE couples to KAWs at this frequency below ω_0, the outgoing KAWs would play a role in TAE damping. This damping is called "radiative damping" of TAE. This type of TAE damping is not dissipative as the wave energy of TAE is transformed into radially outgoing energy flux of KAWs. The value of the radiative damping for TAE with frequency (6.26) below ω_0 is calculated in Appendix C as [6.15,6.16]:

$$\frac{\gamma}{\omega} = -\frac{\pi^2 \varepsilon S^2}{8}\cdot\exp\left(-\frac{\pi^3 S^2}{8\sqrt{2}\lambda}\right), \quad S^2 \ll 1,$$

$$\frac{\gamma}{\omega} = -\frac{\pi^{3/2}}{48}\left(2.5\frac{r}{R}\right)\frac{\lambda^a}{S^4}\exp\left(-\frac{2^{5/2}}{3\lambda}\right), \quad a \cong \frac{\pi^2\sqrt{2}}{144 S^4 \lambda}, \quad S \geq 1. \tag{6.30}$$

where

$$\lambda^2 \equiv \frac{16m^2\rho_i^2}{r^2}\cdot\frac{S^2}{\varepsilon^3}\cdot\left(\frac{3}{4}+\frac{T_e}{T_i}\left(1-i\delta_e\right)\right) \tag{6.31}$$

is the key non-ideal parameter associated with the finite Larmor radius in KAW description.

Above the central frequency, ω_0, KAWs with poloidal harmonics m and $m+1$ overlap. They initially propagate radially towards each other, then form an interference pattern around the TAE-gap localisation region, as shown in Figure 6.4. The overlap and interference of the two KAWs results in two very important physical consequences:

i. the outgoing KAW energy flux that escapes from the KAW interference zone is negligibly small and gives a correspondingly low radiative damping rate of any TAE with frequency above ω_0 [6.15,6.16]
ii. KAWs with frequencies just above the TAE-gap, that is, $g \approx +1+\delta g, 0 < \delta g \ll 1$, propagate towards each other and form a discrete spectrum of standing waves instead of the Alfvén continuum existing in this frequency range within the ideal MHD framework. Such standing waves of KAWs in the narrow radial region surrounding the TAE-gap are, by analogy with (6.15), inner solutions of some modes. If the matching procedure can be performed between these inner solutions and the MHD-type outer solution (6.24), modes with a narrow, but essentially kinetic region surrounding the TAE-gap, could be found. Such modes are called kinetic TAE (KTAE) [6.11,6.13,6.17] Eigenfrequencies of these modes are calculated in the Appendix C. They form a discrete spectrum determined by the non-ideal parameter λ. In the normalised frequency units given by (6.13), the discrete spectrum of KTAEs is:

$$g = 1+\frac{\lambda}{\sqrt{2}}\cdot(4p+1)\cdot\left(1+\frac{S\sqrt{\lambda}}{8\sqrt{2}}\cdot\frac{\Gamma\left(p+\frac{1}{2}\right)}{\Gamma(p+1)}\right), \quad p = 0,1,2\ldots; \, S^2 \ll 1. \tag{6.32}$$

Because KTAEs have frequencies above ω_0, their radiative damping is very small [6.15]:

$$\frac{\gamma}{\omega} = -\frac{\lambda^{\frac{3}{2}}\varepsilon S}{2^{5/4}}\cdot\exp\left(-\frac{2^{5/2}}{3\lambda}\right), \quad S^2 \ll 1, \tag{6.33}$$

and becomes significant only at $\lambda \sim 1$.

Finally, we underline the applicability conditions of Figure 6.4 and Eq. (6.29). One can see in Figure 6.4 that an interference pattern could be also formed at low frequencies of KAWs around $k_{\|m} = 0$ and $k_{\|m+1} = 0$. In these cases, KAWs of the same poloidal harmonic, say, m in the vicinity of $k_{\|m} = 0$, propagate towards each other and may form a discrete spectrum. However, the first-order Larmor radius approximation (6.29) is insufficient for KAW description in this region because $k_r \to \infty$ as $k_{\|m} \to 0$ for any finite frequency ω. A full-order Larmor radius must be considered for KAWs according to Ref. [6.18]. It was found in Ref. [6.18] in a slab geometry that the effects of Larmor radius introduced only to the first order and without dissipation removes the continuous MHD spectrum and the singular modes and replaces them with a discrete spectrum of real frequencies. However, when the full ion Larmor radius response is incorporated in a non-dissipative plasma model, this discrete spectrum disappears and a continuum reappears.

However, the full-order Larmor radius response is not essential for KTAE or for on-axis kinetic modes described by Rosenbluth and Rutherford in Ref. [6.2] as long as the positions of $k_{\|m} = 0$ remain well outside the mode localisation region.

6.3 TAE INTERACTION WITH ENERGETIC PARTICLES AND THERMAL PLASMA, TAE STABILITY

Now we consider how the particle-to-wave power transfer works for TAE in the presence of charged energetic particle population. First, we clarify the components of the field perturbations associated with TAE, for which solutions (6.15), (6.16), and (6.20), (6.21) were found. The incompressible ideal MHD approach used for TAE above allows neither parallel electric nor parallel magnetic perturbations, $\delta E_\| = 0$, $\delta B_\| = 0$. The shear Alfvén waves with perturbed scalar φ and vector δA potentials satisfy then:

$$\delta B_\| \equiv \delta B \cdot B_0 / B_0 = 0 \to \delta A = \delta A_\| \left(B_0 / B_0\right) \equiv \delta A_\| b,$$

$$\delta E_\| \equiv \delta E \cdot \frac{B_0}{B_0} = 0 = -b \cdot \nabla \varphi - \frac{1}{c}\frac{\partial}{\partial t}\delta A_\| \to k_\| \varphi = \frac{\omega}{c}\delta A_\|.$$

We express perturbed m-th perpendicular electric and magnetic fields via φ_m:

$$\delta E_r = -\frac{\partial\varphi_m}{\partial r}\exp\left(i\left[n\zeta - m\vartheta - \omega t\right]\right) + c.c. \tag{6.34}$$

$$\delta E_\vartheta = \frac{im}{r}\varphi_m\exp\left(i\left[n\zeta - m\vartheta - \omega t\right]\right) + c.c. \tag{6.35}$$

$$\delta B_r = -\frac{k_\| c}{\omega}\cdot\frac{im}{r}\varphi_m\exp\left(i\left[n\zeta - m\vartheta - \omega t\right]\right) + c.c. \tag{6.36}$$

$$\delta B_\vartheta = -\frac{k_\| c}{\omega}\cdot\frac{\partial\varphi_m}{\partial r}\exp\left(i\left[n\zeta - m\vartheta - \omega t\right]\right) + c.c. \tag{6.37}$$

We introduce an additional factor corresponding to an exponential growth or damping rate of the mode amplitude due to the particle-to-wave power transfer at an early linear phase of the wave-particle interaction,

$$\varphi(r,\vartheta,\zeta,t) \to \varphi(r,\vartheta,\zeta,t) \cdot \exp(\gamma t). \tag{6.38}$$

The net exponential growth/damping rate of TAE amplitude is determined by

$$\gamma = \frac{P_\alpha - P_d}{2 \cdot \delta W} \equiv \gamma_\alpha - \gamma_d, \tag{6.39}$$

where P_α is the power transfer from charged energetic particles to the wave, P_d is the power absorbed by the background plasma, and δW is the wave energy given by the sum of field energy and kinetic energy due to $\delta E \times B_0$ drift:

$$\delta W = \int d\zeta d\vartheta dr \cdot Rr \left(\frac{\delta B_\vartheta^2}{8\pi} + \frac{\omega^2}{k_\parallel^2 c^2} \cdot \frac{\delta B_\vartheta^2}{8\pi} \right). \tag{6.40}$$

The mode is linearly stable when $\gamma \le 0$.

6.3.1 POWER TRANSFER FROM ENERGETIC PARTICLES TO TAE

For considering energetic particle drive, we take an example of burning plasma in ITER baseline scenario with current $I_P = 15$ MA, $R_0 = 621$ cm, $a = 200$ cm, and $B_0 = 5.3$ T, $n_e(0) \approx 10^{20}$ m^{-3}. For TAE, 1 MeV deuterium beam ions and alpha particles we obtain $V_A = 7 \cdot 10^8$ cm/s $< V_{\text{beam}}(t = 0) = 10^9$ cm/s $< V_\alpha(t = 0) = 1.3 \cdot 10^9$ cm/s, so that both types of the energetic ions during their slowing-down process could enter resonance $V_\parallel = V_A$.

Because TAE has no parallel electric field, $\delta E_\parallel = 0$, the power transfer from a resonant alpha particle to the wave comes in the guiding centre approach from the alpha particle orbit drifting across the TAE magnetic flux surface:

$$P_\alpha = \delta E \cdot J_\alpha = -e_\alpha V_d \cdot \delta E, \tag{6.41}$$

where the unperturbed guiding centre drift velocity is

$$V_d = -V_d (\sin \vartheta \cdot e_r + \cos \vartheta \cdot e_\vartheta), \tag{6.42}$$

$$V_d = \frac{V_\perp^2 / 2 + V_\parallel^2}{\omega_{B\alpha} R}, \omega_{B\alpha} = \frac{e_\alpha B_0}{m_\alpha c}. \tag{6.43}$$

Figure 6.5 shows a two-dimensional projection of passing or trapped drift resonant alpha particle orbit moving from point A to point B across the radial structure of TAE mode with electric field δE.

When a resonant ion moves radially across TAE from point A to point B, it crosses the electrostatic potential associated with TAE. The mode and the ion exchange energy $e\Delta\varphi$ during the radial drift motion is shown in Figure 6.5.

We can see here that the particle-to-wave power transfer arises because TAE is attached to magnetic flux surface, while the drift orbit moves across in radius and is displaced from the magnetic surface. Furthermore, it is easy to understand from Figure 6.6 that the particle-to-wave power transfer depends on the ratio between the radial width of the orbit and the radial width of TAE. In fact,

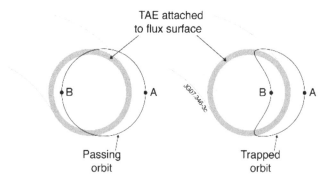

FIGURE 6.5 Schematic geometry of the interacting TAE attached to the magnetic flux surface, and passing (left) or trapped (right) drift orbits of the interacting resonant particles displaced from the magnetic surfaces.

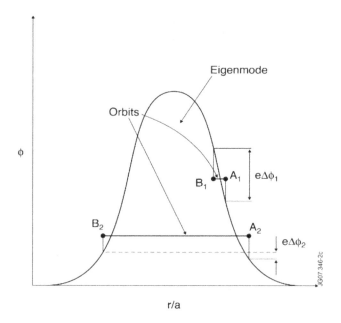

FIGURE 6.6 Schematic radial projection of the two-dimensional plot in Figure 6.5 showing the drift orbit crossing one harmonic of the electrostatic potential associated with TAE.

there are two characteristic radial widths of TAE. The one determined by the mode structure in the inner layer (6.15) and (6.16) where toroidal coupling matters is

$$\Delta_{\text{in}} \sim \varepsilon \frac{r_{\text{TAE}}}{m}. \tag{6.44}$$

The second characteristic TAE width is determined by the outer mode structure (Eqs. 6.20 and 6.21):

$$\Delta_{\text{out}} \sim r_{\text{TAE}} / m. \tag{6.45}$$

Figure 6.7 shows qualitatively how the growth rate of energetic particle-driven TAE increases linearly at small $\Delta_O / \Delta_m \equiv \Delta_O \cdot m / r_{\text{TAE}}$ corresponding to the A_1–B_1 orbit in Figure 6.6. The growth

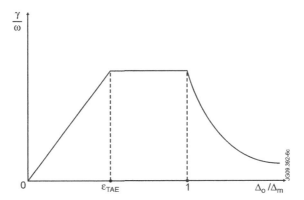

FIGURE 6.7 Qualitative dependence of the energetic particle drive for TAE as a function of the ratio between the energetic particle drift orbit width and the radial width of TAE.

rate saturates when the orbit becomes comparable to the inner width of TAE [6.19], and decreases at the orbit size much larger than the outer width of TAE (this case corresponds to the A_2–B_2 orbit in Figure 6.5 [6.11]). The highest energetic ion drive is achieved at

$$m \approx nq \approx r_{\text{TAE}} / \Delta_O,\tag{6.46}$$

when the maximum is achieved in the value $|e_\alpha \Delta \varphi|$. at the drift orbit width equal to the mode width.

We can extend the particle-to-wave power transfer from the single particle consideration to a distribution function of energetic particles. The particle-to-wave power transfer for the whole population of energetic particles takes the form

$$P_\alpha = \int \mathrm{d}\zeta \mathrm{d}\vartheta \mathrm{d}r \cdot Rr \int \mathrm{d}^3v \left(-e_\alpha V_d \cdot \delta E\right) \cdot f,\tag{6.47}$$

where f is the linear perturbed distribution function of energetic particles.

For calculating (6.47), we assume for simplicity that [6.19]

$$\omega \ll \omega_{*\alpha} = -\frac{m}{r} \cdot \frac{cE_\alpha}{e_\alpha B_0} \cdot \left(\frac{1}{p_\alpha}\frac{\mathrm{d}p_\alpha}{\mathrm{d}r}\right),\tag{6.48}$$

and consider passing ions for which $\dot{\phi} = v_\parallel / R, \dot{\vartheta} = v_\parallel / qR$. The drift kinetic equation for energetic ions takes the following form:

$$\frac{\partial f}{\partial t} - V_d \sin\vartheta \frac{\partial f}{\partial r} - V_d \cos\vartheta \frac{1}{r}\frac{\partial f}{\partial \vartheta} + \dot{\phi}\frac{\partial f}{\partial \phi} + \dot{\vartheta}\frac{\partial f}{\partial \vartheta} =$$

$$= \frac{c}{B}\left(1 - \frac{k_\parallel v_\parallel}{\omega}\right)\frac{1}{r}\frac{\partial f_0}{\partial r}\left(-im\varphi_m + \frac{\partial \varphi_m}{\partial \vartheta}\right)\exp\left(i\Psi_m\right) + c.c.,\tag{6.49}$$

where $\Psi \equiv n\phi - m\vartheta - \omega t$.

To represent the electrostatic potential φ_m of TAE in the reference frame associated with the drift orbits of passing ions, we change variables $(r, \vartheta) \to (\bar{r}, \vartheta)$ meaning the transform to the orbit reference frame:

$$\bar{r} = r - \Delta_O \cos\vartheta,\tag{6.50}$$

$$\bar{\vartheta} = \vartheta + \frac{\Delta_O}{r} \sin \vartheta \approx \vartheta. \tag{6.51}$$

This change in the variables gives

$$\frac{\partial}{\partial r} = \frac{\partial}{\partial \bar{r}}; \frac{\partial}{\partial \vartheta} = \Delta_O \sin \bar{\vartheta} \frac{\partial}{\partial \bar{r}} + \frac{\partial}{\partial \bar{\vartheta}},$$

so Eq. (6.49) transforms to

$$\frac{\partial f}{\partial t} + \dot{\phi} \frac{\partial f}{\partial \phi} + \dot{\vartheta} \frac{\partial f}{\partial \vartheta} = \frac{c}{B}\left(1 - \frac{k_\parallel v_\parallel}{\omega}\right)\frac{1}{r}\frac{\partial f_0}{\partial r}\left(-im\varphi_m + \frac{\partial \varphi_m}{\partial \vartheta}\right)\exp(i\Psi_m) + c.c. \tag{6.52}$$

where the electrostatic potential of TAE seen by the energetic ion along its orbit depends on the poloidal angle ϑ and can be represented as a Fourier series in ϑ:

$$\varphi_m\left(\bar{r} + \Delta_O \cos \vartheta\right) = \sum_{l=0}^{\infty} \varphi_{m,l} \cos l\vartheta. \tag{6.53}$$

The wave-particle resonance condition for the passing ions has the form

$$(\Psi_m) \equiv n\dot{\phi} - (m+l)\dot{\vartheta} - \omega = 0, \tag{6.54}$$

so the resonance contribution from the kinetic equation (6.52) takes the form:

$$f = -\frac{c}{2B\omega}\frac{1}{r}\frac{\partial f_0}{\partial r}\sum_l (m+l)\left(\dot{\Psi}_m \varphi_{m,l} + \dot{\Psi}_{m+1}\varphi_{m+1,l}\right)(-i\pi)\delta\left(\omega - n\dot{\phi} + (m+l)\dot{\vartheta}\right)\exp\left[i(\Psi_m - l\vartheta)\right] + c.c.$$

Integration of (6.47) for this perturbed distribution function and for strongly passing, $\frac{V_\parallel}{V} = 1$, beam distribution function f_0 gives

$$\frac{\gamma_\alpha}{\omega} = -\frac{15}{16\pi}q^2 r_{AE}\frac{d\beta_{beam}}{dr} \cdot \frac{V_A^2}{V_0^2} \cdot \frac{I(D)}{D}, \tag{6.55}$$

where the finite drift orbit parameter D is given by

$$D = \frac{qmV_A}{r_{AE}\omega_{Bb}}, \tag{6.56}$$

estimated at the resonant velocity $V_\parallel = V_A$, and the coupling integral is shown in Figure 6.8.

Note that the resonance condition (6.54) takes the following form for TAE:

$$V_\parallel = \frac{V_A}{|1 - 2l|}, \tag{6.57}$$

where l = integer describes the wave-particle resonances. In particular, the principal resonance,

$$V_\parallel = V_A \tag{6.58}$$

corresponds to $l = 0$,

and the strongest side-band resonance,

$$V_\parallel = \frac{V_A}{3} \tag{6.59}$$

corresponds to $l = -1$.

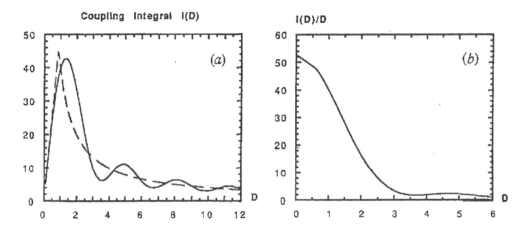

FIGURE 6.8 Coupling integral for passing beam particle principal resonances: (a) asymptotic expression (broken curve) and numerical result (full curve); (b) function $I(D)/D$ used in Eq. (6.55).

For isotropic distribution function of fusion-generated alpha particles, the alpha particle drive for TAE is estimated to be

$$\frac{\gamma_\alpha}{\omega} = -\frac{3}{32\pi}\varepsilon q \frac{d\beta_\alpha}{dr}\frac{V_A^2}{V_0^2}\int_{V_A}^{V_0} dV\frac{V_A^2+V^2}{V^2}\cdot\frac{I(D)}{D}, \tag{6.60}$$

where the finite drift orbit parameter D for passing alpha particles resonating with TAE via the $V_{\|\alpha} = V_A$ resonance is given by

$$D = \frac{qm\left(V_A^2+V^2\right)}{2V_A r_{AE}\omega_{B\alpha}} \tag{6.61}$$

The drive theory could be extended to a general case of wave-particle resonances including trapped particles and the distribution with gradients in both velocity space and in the real space. The description of a guiding centre distribution function in a five-dimensional phase space could be significantly simplified with a proper choice of variables. In tokamaks with toroidally axisymmetric magnetic field, three invariants of the unperturbed particle orbits are conserved: particle energy $E \equiv m_\alpha V^2 / 2$, magnetic moment $\mu \equiv m_\alpha V_\perp^2 / (2B)$, and the toroidal angular momentum,

$$P_\phi \equiv -(e_\alpha / c)\Psi(r) + m_\alpha R V_\phi \tag{6.62}$$

where $\Psi(r)$ is the poloidal flux, $V, V_\perp, V_\phi$ are total, perpendicular (to the magnetic field), and toroidal velocities of the particle, e_α, m_α are the charge and mass of the particle, and R is the major radius. It is convenient to use the variables (E, P_ϕ, μ) for describing the energetic particle distribution function, $f_\alpha(E, P_\phi, \mu)$, and the characteristic toroidal, $\omega_\phi(E, P_\phi, \mu)$, and poloidal, $\omega_\vartheta(E, P_\phi, \mu)$, orbit frequencies of the particles. The resonance condition between TAE with frequency ω and toroidal mode number n and the energetic particles takes the following form in the general case:

$$\Omega \equiv \left(\dot\Psi_m\right) = n\dot\phi - (m+l)\dot\vartheta - \omega = n\omega_\phi\left(E, P_\phi, \mu\right) - p\omega_\vartheta\left(E, P_\phi, \mu\right) - \omega = 0,$$

$$p = 0, 1, 2... \tag{6.63}$$

The wave growth rate in the general form has the contributions from the gradients of the energetic particle distribution in both radial and energy space:

$$\gamma_L \propto \int d^3\boldsymbol{r} \int dP_\phi dE \left(-e_\alpha V_d \delta E\right) \delta(\Omega) \left(\omega \frac{\partial}{\partial E} + n \frac{\partial}{\partial P_\phi}\right) f_\alpha\left(E, P_\phi, \mu\right). \tag{6.64}$$

6.3.2 TAE Damping due to Thermal Ions

Due to the side-band resonance (6.59), the exponential tail of the Maxwellian distribution of thermal ions may interact with TAE. The efficiency of the resonance $V_\parallel = \dfrac{V_A}{3}$ is significantly lower than the efficiency of the principal resonance $V_\parallel = V_A$ between TAE and energetic ions. However, the density of thermal ions exceeds density of energetic ions by several orders of magnitude, and hence, the power transfer between thermal ions and TAE may be comparable to that between energetic ions and TAE [6.20].

Thermal ion species burning ITER plasmas are of four main types:

$$\text{D:} \quad m_D = 2m_H, \quad e_D = e_H;$$

$$\text{T:} \quad m_T = 3m_H, \quad e_T = e_H;$$

$$\text{He:} \quad m_{He} = 4m_H, \quad e_{He} = 2e_H;$$

$$\text{Be:} \quad m_{Be} = 9m_H, \quad e_{Be} = 4e_H;$$

where m_H and e_H are the mass and charge of hydrogen ion (proton). For these ions, the quasi-neutrality condition is:

$$n_e = n_D + n_T + 2n_{He} + 4n_{Be}, \tag{6.65}$$

and the plasma depletion due to the impurity ions is as follows:

$$(n_D + n_T) / n_e = 1 - \mu. \tag{6.66}$$

Here, we consider the depletion factor in the range

$$0.08 \leq \mu \ll 1, \tag{6.67}$$

where the lower boundary is determined by Be impurity ions coming from Be wall in ITER and always present in the plasma at the level of ~2%. Another major contributor to the depletion is He ash population, the density of which is determined by the intensity of DT fusion and the efficiency of He ash removal by He transport and pump-out. The D:T mixture in burning plasmas is supposed to be close to 50:50, so we assume

$$n_T / n_D = 1 - \delta, \quad |\delta| \ll 1. \tag{6.68}$$

It follows then that the ion densities are related to the electron density via:

$$\frac{n_D}{n_e} = \frac{1-\mu}{2-\delta}; \frac{n_T}{n_e} = \frac{(1-\delta)\cdot(1-\mu)}{2-\delta}; \frac{n_{He}}{n_e} = \frac{\mu}{2} - 0.04; \tag{6.69}$$

and Be density is 2%. The value of Alfvén velocity in such plasma is

$$V_A = \frac{B_0}{\sqrt{4\pi m_H \left(2n_D + 3n_T + 4n_{He} + 9n_{Be}\right)}} = \frac{B_0}{\sqrt{4\pi m_H n_e \cdot 2.43 \cdot \left(1 - 0.2\mu - 0.08\delta\right)}}, \quad (6.70)$$

showing a higher sensitivity to the plasma depletion than to the deviation in the D:T mixture from the optimum value 50:50.

The drift frequency of thermal ions is much lower than TAE frequency, $|\omega_{*i} / \omega_{TAE}| \ll 1$, so we can neglect the drift ion effects and consider only the contribution due to the term $\omega \partial F_i / \partial E$ in (6.64) to obtain the estimate of thermal Dion Landau damping [6.15]:

$$\frac{\gamma_D}{\omega} \cong -\frac{\sqrt{\pi}}{4} q^2 \beta_D x_D \left(1 + \left(1 + 2\tau_D + 2x_D^2\right)^2\right) \exp\left(-x_D^2\right). \quad (6.71)$$

Here, the term with $\tau_D = T_e / T_D$ is caused by finite parallel electric field of TAE, and the plasma was assumed with D ions having the Maxwellian distribution function and thermal velocity V_{TD}:

$$x_D = \frac{V_A}{3V_{TD}}. \quad (6.72)$$

Note here that the ion Landau damping does not depend on the mode number.

For the plasma consisting of a D-T mix of ions the ion Landau damping is a sum of the contributions from D ions and T ions,

$$\gamma_i = \gamma_D + \gamma_T, \quad (6.73)$$

where damping due to D ions is determined by (6.71) with D ion density and temperature, and γ_T is given by a similar expression with index T instead of D. Combining the damping effects due to D and T thermal ions gives

$$\frac{\gamma_i}{\omega} = \frac{\gamma_D}{\omega} \cdot \left[1 + \frac{n_T}{n_D} \sqrt{\frac{3}{2}} \left(\frac{1 + x_D^2}{1 + (2/3)x_D^2}\right)^2 \exp\left(-\frac{x_D^2}{2}\right)\right], \quad (6.74)$$

so the relative contribution of T ions could be easily assessed.

6.3.3 TAE Damping due to Thermal Electrons

For ITER baseline scenario, we obtain the following ordering:

$$V_A = 7 \times 10^8 \, \text{cm/s} \ll V_{Te} \sim 9.3 \times 10^9 \, \text{cm/s}. \quad (6.75)$$

On the other hand, for an effective electron Landau damping, the principal resonance condition should be met,

$$V_{\|e} = V_A. \quad (6.76)$$

Combining (6.75) and (6.76), we obtain for thermal electrons

$$V_{\|Te} \ll V_{Te}. \quad (6.77)$$

This inequality could only be possible if

$$V_{\|Te} \ll V_{\perp Te}, \tag{6.78}$$

that is, the majority of electrons satisfying (6.76) are trapped. For trapped electrons, the main mechanism of electron damping is associated with Coulomb scattering of the electrons near the trapped-passing boundary [6.14,6.17,6.21], which gives

$$\frac{\gamma_e}{\omega} \cong -\left(\frac{\nu_e}{\omega}\right)^{1/2}\left(4\beta_e q^2 + 0.44\left(\frac{\rho_S}{\Delta_{TAE}}\right)^2\right) \cdot \left(\ln\left[16\left(\frac{\varepsilon\omega}{2\nu_e}\right)^{1/2}\right]\right)^{-3/2} \tag{6.79}$$

This damping effect depends on the m-number (via Δ_{TAE}). In particular, this effect could effectively absorb outgoing KAW waves when $k_r \to \infty$.

Although the electron Landau damping of TAE is small at the position of the TAE-gap where the mode phase velocity is Alfvén velocity, the phase velocity, $\omega/k_{\|m}$, can increase significantly away from the TAE-gap as the eigenfrequency is fixed while the parallel wave-vector depends on the radius, $k_{\|m} = k_{\|m}(r)$. Electron Landau damping may contribute then [6.22], but it affects the tails of the mode structure (6.24), which are small.

6.3.3   CONTINUUM DAMPING OF TAE

Finally, some continuum damping of TAE may still exist if the TAE frequency crosses the Alfvén continuum at a certain point in radius. This damping could vary from zero in the case of the so-called "open" TAE-gap (no crossing exists) to a high value when the TAE-gap is closed and the continuum crossing point in radius is close to the TAE-gap. This effect was experimentally validated on JET, where the TAE damping measured with the external TAE antenna varied from 0.6% in the open TAE-gap case, to a very high value of 5% in the case of a closed TAE-gap [6.23]. In a limiting case of very high mode numbers and when the TAE-gap is closed, the expression for the continuum damping was obtained in Refs. [6.24,6.25]:

$$\frac{\gamma_{cont}}{\omega} \cong -0.8\frac{S^2}{m^{3/2}\sqrt{2.5r/R}}. \tag{6.80}$$

REFERENCES

1. A.B. Mikhailovskii, *Sov. Phys. JETP* **41** (1975) 980.
2 M.N. Rosenbluth and P.H. Rutherford, *Phys. Rev. Lett.* **34** (1975) 1428.
3. G.Y. Fu, PhD Thesis, The University of Texas at Austin (1988).
4. C.Z. Cheng et al., *Ann. Phys. (N.Y.)* **161** (1984) 21.
5. G.Y. Fu and J.W. Van Dam., *Phys. Fluids* **B1** (1989) 1919.
6. L.Chen, *Theory of fusion plasma* (Proc. Varenna Int. Workshop), edited by J. Vaclavik, F. Troyon, and E. Sindoni, Societa Italiana di Fisica-Editrice Compositori, Bologna (1989) p.327.
7. K.L. Wong et al., *Phys. Rev. Lett.* **66** (1991) 1874.
8. W.W. Heidbrink et al., *Nucl. Fusion* **31** (1991) 1635.
9. R. Nazikian et al., *Phys. Rev. Lett.* **78** (1997) 2976.
10. H.L. Berk et al., *Phys. Fluids* **B4** (1992) 1806.
11. B.N. Breizman and S.E. Sharapov, *Plasma Phys. Control. Fusion* **37** (1995) 1057.
12. A.B. Mikhailovskii and S.E. Sharapov, *Spectrum of Ideal Ellipticity-Induced Alfven Eigenmodes in a Low-Shear Tokamak*, Proceed. of the 5th IAEA TCM on Alpha Particles in Fusion Research (JET Joint Undertaking, Abingdon, Oxfordshire, UK 1997).
13. R.R. Mett and S.M. Mahajan, *Phys. Fluids* **B4** (1992) 2885.
14. J. Candy and M.N. Rosenbluth, *Nucl. Fusion* **35** (1995) 1069.

15. J.W. Connor et al., *Non-ideal Effects on TAE Stability*, Proceed. of the 21st EPS Conference on Control. Fusion and Plasma Phys., edited by E. Joffrin, P. Platz, and P.E. Stott, (Montpellier, Vol. 18B, 27 June 1994), p. 616.

16. R. Nyqvist and S.E. Sharapov, *Phys. Plasmas* **19** (2012) 082517.

17. J. Candy and M.N. Rosenbluth, *Phys. Plasmas* **1** (1994) 356.

18. J.W. Connor et al., *Phys. Fluids* **26** (1983)158.

19. H.L. Berk et al., *Phys. Lett.* **A162** (1992) 475.

20. R. Betti and J.P. Freidberg, *Phys. Fluids* **B3** (1991) 1865.

21. N.N. Gorelenkov and S.E. Sharapov, *Physica Scripta* **45** (1992) 163.

22. J. Candy, *Plasma Phys. Control. Fusion* **38** (1996) 795.

23. A. Fasoli et al., *Phys. Rev. Lett.* **75** (1995) 645.

24. M.N. Rosenbluth et al., *Phys. Rev. Lett.* **68** (1992) 596.

25. F. Zonca and L. Chen, *Phys. Rev. Lett.* **68** (1992) 592.

.

7 Experimental Studies of Alfvén Eigenmodes

The usual theoretical approach to understanding energetic particle-driven instabilities is that the number of energetic particles is a minor perturbation to the background plasma. For electromagnetic Alfvén eigenmodes, this condition implies significantly lower pressure of the energetic particles than that of the background plasma, $\beta_{hot} \ll \beta_{therm}$. In this framework, the real part of the mode frequency, ω, is determined by the thermal plasma properties, and the small negative imaginary part of frequency is the mode damping, γ_d, due to the thermal plasma. The energetic ions resonating with the mode only affect the small imaginary part of the mode frequency providing the growth rate γ_L and exciting the mode if the growth rate exceeds the mode damping, $\gamma_L > |\gamma_d|$.

Experimentally, the characteristic parameters of weakly damped AEs are investigated in two different ways depending on the presence of energetic particles. In the absence of energetic particles, the very presence of weakly damped AEs, the real parts of the AE frequencies, and the AE damping rates due to thermal plasma are studied from a resonance plasma response to an electromagnetic wave launched with an external antenna at AE frequencies. The resonances are searched by probing the plasma with an externally launched wave of a frequency swept across the expected frequency range of the AEs. By measuring the plasma response to the wave launched, resonances with high quality factor $Q = \omega/\gamma = 10^3 - 10^2$ can be detected at the frequencies of weakly damped AEs. This concept is similar to the numerical experiment [7.1,7.2] that discovered GAE (see Chapter 5). The technique of probing plasma with external antenna and identifying AE resonances in plasma response has been extensively used in JET for studies of stable TAEs since the 1990s [7.3]. These JET studies will be described in Section 7.1.

When energetic particles have sufficiently high enough pressure gradient so that the energetic particle drive exceeds the damping of an AE, energetic particles excite the AE. The unstable AEs in the linear phase increase exponentially in amplitude, and exhibit one of the two main non-linear scenarios involving the AE amplitude and frequency [7.4]. The first scenario has the mode frequency locked (FL) to plasma equilibrium, that is, the frequency of the AE at its non-linear phase is similar to the linear AE frequency as determined by the thermal plasma at the same time. In the case of FL modes, the mode amplitude is nearly constant or varies on a relatively slow time scale of plasma equilibrium changes, such as variations in plasma density or safety factor. The second main scenario has the mode frequency sweeping (FS) on a very short time scale, much faster than equilibrium can change, with the mode amplitude bursting quasi-periodically. Figure 7.1 presents a typical example of FL AEs on the JET tokamak with ICRH-accelerated ions [7.5], while Figure 7.2 presents FS Alfvén instability on the JT-60U tokamak with negative NBI [7.6].

In the case shown in Figure 7.1, the Alfvén perturbations exhibit a discrete spectrum of TAEs with different toroidal mode numbers n and frequencies, which are determined by the bulk plasma equilibrium. The observed slow variation in TAE frequencies is caused by an increase in plasma mass density throughout the time window shown. The mass density $\rho(t)$ of the plasma increases due to the beam fuelling of the discharge, and TAE frequency evolves in accordance with the Alfvén scaling $V_A \propto B / \sqrt{\rho(t)}$. Amplitudes of these TAEs saturate and remain nearly constant. TAEs of FL type will be considered in Section 7.2.

In contrast to the FL scenario, Figure 7.2 shows an FS Alfvén instability on JT-60U, with frequency of the Alfvénic perturbations sweeping up and down on a time scale much shorter than the time scale of plasma equilibrium. Amplitude of this FS instability exhibits bursts, and the mode

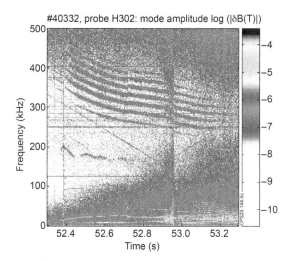

FIGURE 7.1 Magnetic spectrogram of ICRH-driven TAEs of FL type with frequencies ~250–500 kHz in JET discharge #40332 with ICRH.

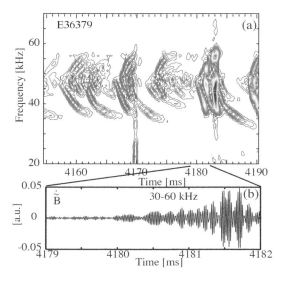

FIGURE 7.2 (a) Magnetic spectrogram of beam-driven Alfvén instabilities of FS type in JT-60U, $B = 1.2$ T, ENBI = 360 keV. (b) Raw Mirnov coil signal.

frequency sweeps by ~10%–40% during every burst. Phenomenology and more examples of such FS modes are considered in Section 7.3, and the relevant theoretical models are presented in Chapter 8.

We underline here that the FL and FS scenarios of energetic particle-driven instabilities require different concepts in their modelling. Namely, the frequency and structure of unstable modes in the FL scenario remain nearly the same as those of the corresponding linear AEs throughout the entire linear and non-linear evolution of the mode. In this case, the modes are determined by thermal plasma equilibrium, while the energetic particles determine mode growth rate only. Experimentally measured structure of such modes remains similar to the linear mode structure, and for modelling such modes one can use the eigenmode structure and eigenfrequency computed with a linear MHD code.

In the FS scenario, contribution of energetic particles to mode frequency is as essential as the bulk plasma contribution and cannot be found perturbatively. In the non-linear phase, when the unstable mode re-distributes the energetic particles contributing to the mode frequency, the mode adapts its own frequency to the re-distributed population of energetic ions. The characteristic time scale of the energetic particle re-distribution is comparable to the inverse growth rate, and hence, the mode frequency determined by energetic particles changes together with energetic particles on the time scale of the inverse growth rate. Modelling of such "non-perturbative" modes must include fully non-linear description of wave-particle interaction and, for some modes, MHD non-linearities as well.

The transport of energetic particles caused by AEs with FL differ considerably from the one caused by AEs with the FS regime. In particular, the bursting FS modes can lead to a flux of fast ion losses with high peak values, while a time-averaged flux of these lost ions could be low. Next, FS modes affect energetic particles in a broader phase space area due to the swept frequency involved in the wave-particle resonance. Finally, a rather significant convective transport of energetic particles could be caused by an FS mode, while a global stochastic diffusion of energetic particle orbits is the most significant transport effect for FL modes.

7.1 PROBING STABLE ALFVÉN EIGENMODES WITH EXTERNAL TAE ANTENNA

The primary aim of theoretical and experimental studies of AEs is to describe their dispersion properties and to quantify the main mechanisms that damp or drive AE instabilities. Results of these studies can be used then to predict with a higher confidence the next-step operating regimes of DT plasmas in machines such as ITER. To investigate the spectrum of stable weakly damped AEs and their damping rates separately from the mode drive, a dedicated active diagnostic system was developed on JET [7.3,7.7,7.8]. This active TAE probing technique uses an external antenna for launching an electromagnetic wave into the plasma, with the wave frequency swept across the frequency range of AEs. A synchronous detection technique measures the plasma response to the wave launched. On JET, saddle coils were used initially as the external TAE antenna [7.3]. Saddle coils were chosen for this role owing to their extended structure covering the entire plasma in toroidal direction, thus allowing the launch of waves with parallel wave-vectors, $k_\parallel \sim 1/qR$, relevant to TAEs. Figure 7.3 shows an example of the launched probing wave swept up and down across the TAE frequency, and measuring the plasma response as perturbed magnetic fields with synchronous detection using Mirnov coils (description of Mirnov coils is given in Appendix D). The peak response in the synchronous detection is associated with high-quality TAE resonance (low damping rate). The frequency of the resonance corresponds to TAE with the specific toroidal mode number $n = 1$ launched by the antenna, while the damping rate of the mode is characterised by the width of the resonance. The temporal evolution of the frequency and damping rate of TAE seen in Figure 7.3 are caused by the variation in the plasma parameters.

Each individual TAE resonance seen in the magnetic probe signal is represented in a complex plane, as shown in Figure 7.4, and a best-fit routine is used to assess the width of the resonance that provides the information on the mode damping. In the case of Figure 7.4, two upper saddle coils were used in phase and 180° apart toroidally.

One of the first scientific results obtained with the active TAE diagnostics was the experimental validation of the continuum damping of a TAE depending upon the alignment of TAE-gaps. Because the centre of the TAE-gap is determined by frequency $f_{TAE} = V_A/(4\pi q R_0)$, the radially separated TAE-gaps corresponding to different m (and the same n) have frequencies of their centres aligned if the plasma profiles satisfy

$$1/q(r)\left(\rho(r)\right)^{\frac{1}{2}} = \text{const.} \qquad (7.1)$$

For the aligned TAE-gaps discharged with the profiles satisfying (7.1), the frequency of TAE does not cross the Alfvén continuum anywhere in the plasma, and hence, there is no continuum

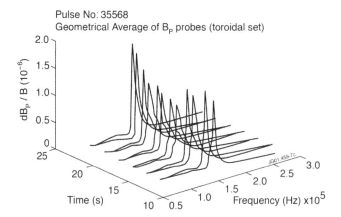

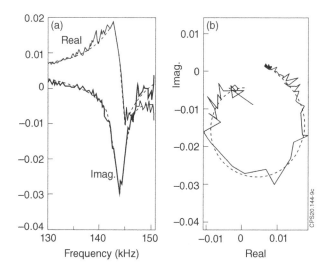

FIGURE 7.3 A TAE resonance is detected by the sweeping frequency of launched wave across estimated TAE frequency and measuring the plasma response with synchronous detection. The measured signal provides information regarding individual TAE frequency and damping.

FIGURE 7.4 Example of a TAE resonance in the ohmic phase of JET shot #31638. Real and imaginary parts (a) and complex plane representation (b) are shown of a magnetic probe signal, normalised to the driving current. The fit with (b) and (a) is also shown, giving $f_{obs} = 144.2 \pm 0.1$ kHz, $\gamma_d/2\pi = 1400 \pm 100$ s^{-1}. $B_{tor} \approx 2.8$ T, $I_p \approx 2.2$ MA, $\langle n_e \rangle \approx 3 \times 10^{19}$ m^{-3}.

damping of TAE. However, if the condition (7.1) is not fulfilled, the eigenfrequency of TAE crosses the Alfvén continuum line at some radius, and the TAE experiences a continuum damping at the crossing point, which may be quite significant, see Section 6.2.3 of Chapter 6 and the References therein.

Figure 7.5 shows the relevant experimental data for two comparison discharges on JET, one of which did not have the TAE-gap alignment condition (7.1) fulfilled, while the other had it fulfilled. The difference in the TAE damping rate for these two cases is about an order of magnitude. In the case of not aligned TAE-gaps, the damping of TAE is very high at approximately 5%, while TAE damping in the case of aligned gaps is only 0.6%. TAE with very high damping corresponds to quite significant continuum damping, as was expected from the plasma profiles not satisfying (7.1), as well as from the modelling [7.7].

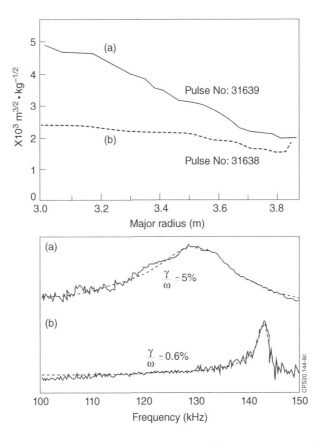

FIGURE 7.5 The relationship between the profile of $1/(q(r)\rho\ (r)^{1/2})$ and the TAE damping. The profile (top) and the raw and fitted frequency responses (bottom) of a normalised magnetic probe signal are shown for two discharges. Excitation peaked at $n=2$ was used for both discharges; measurements were taken in the ohmic phase with similar plasma configuration; $n_e \approx 4 \times 10^{19}\,\text{m}^{-3}$; (a) $B_{\text{tor}} \approx 1.8\,\text{T}$, $I_p \approx 2\,\text{MA}$. (b) $B_{\text{tor}} \approx 2.8\,\text{T}$, $I_p \approx 2.3\,\text{MA}$.

Another important study performed with the JET saddle coils as a TAE antenna was the demonstration of TAE-to-KTAE transition at increasing plasma temperatures [7.9]. Several heating options were explored to observe the structural changes in the TAE spectrum at increasing plasma temperature. Figure 7.6 shows one of these cases, in which higher electron temperature was achieved via ohmic heating of plasma in a discharge with a very significant increase in plasma current from 2 to 4.1 MA.

The TAE antenna was scanning, without tracking any resonances, the frequency from 140 to 260 kHz throughout the discharge. In the early phase of the discharge, when the plasma was relatively cold, only a single TAE resonance was detected, as shown in Figure 7.6a. As the current increases and the electron temperature rises, multiple resonance peaks emerge above TAE frequency identified as a spectrum of KTAEs predicted theoretically in Refs. [6.11,6.13,6.17].

During recent years, the active TAE antenna diagnostic was significantly modified on JET. In particular, this diagnostic technique was enhanced by a digital real-time control system, which allows performing individual TAE resonance tracking. With the use of the tracking system, the AE frequencies and damping rates can be measured with high time resolution (<50 ms) throughout discharges. While the sweeping technique shown in Figure 7.3 provides the opportunity to detect multiple AE resonances within an extended fixed frequency band, the tracking technique adjusts the sweeping frequency range to a narrow width surrounding an individual TAE resonance [7.8]. Figure 7.7 (left) shows an example of magnetic spectrogram with the characteristic zigzag pattern of the launched probing wave swept up and down in frequency. Figure 7.7 (p.85) shows the peak response

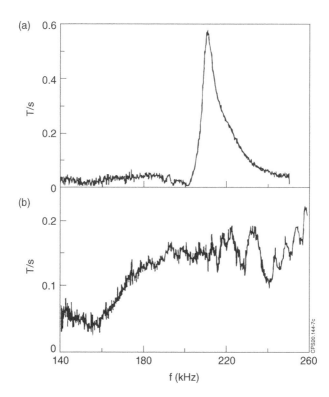

FIGURE 7.6 B_{pol} probe signals for moderate (top) and high plasma current (bottom) in the same discharge #34073. Top: $t = 3.5$ s; $I_p \sim 2$ MA; $B_{tor} \sim 2.5$ T; $\langle n_e \rangle \sim 1.9 \times 10^{19}$ m^{-3}; $T_e \sim 2.2$ keV. The single TAE has $f \sim 210.5$ kHz, $\gamma/\omega \sim 1.4\%$; $f_{TAE}{}^0 \sim 200$ kHz. Later, at $t = 9.5$ s (bottom) multiple peaks appear, with $\Delta f/f \sim 2\%$; $I_p \sim 4.1$ MA; $B_{tor} \sim 2.9$ T; $\langle n_e \rangle \sim 3 \times 10^{19}$ m^{-3}; $T_e \sim 3.2$ keV; $f_{TAE}{}^0 \sim 180$ kHz.

in the synchronous detection corresponding to a TAE resonance detected. Temporal evolution of frequency and damping of TAE, which are determined by varying plasma parameters, are tracked in time by adjusting the sweeping frequency range after every resonance measurement providing high time resolution.

In present-day JET machine, two sets of dedicated TAE antennae are installed at toroidally opposite positions [7.10]. The antennae themselves are much less extended than the saddle coils. However, they can launch waves with a prescribed phase, and can also couple to TAEs with higher mode numbers, $n > 2$. Dedicated active TAE antenna diagnostics were also installed in Alcator C-Mod and MAST tokamaks [7.11].

Finally, JET also employed a technique of probing TAE frequency range with a beat wave from two ICRH antennae launching fast magneto-acoustic waves in the ion cyclotron frequency range of ~ 40–50 MHz [7.12]. In JET, up to four ICRH antennae and four RF generators were individually optimised and used for the best coupling to the plasma. Under these conditions, a small but finite mismatch in the frequencies of the launched waves is always present at the level of 1% or so. This mismatch frequency is close to the TAE frequency range, and so the beat wave between two ICRH antennae matching a TAE frequency could serve as a beat wave antenna inside the plasma suitable for probing the TAE. The JET experiment was successful, but it was found later that the beat wave technique on JET generates perturbations with parallel wave-vectors varying in an uncontrollable manner, very quickly and randomly. This made the technique of TAE probing by the beat wave much less attractive than the external TAE antenna probing.

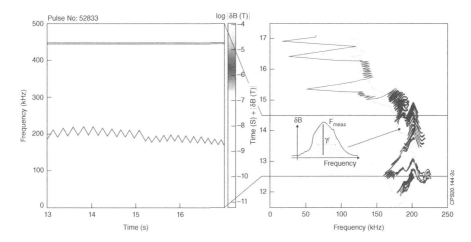

FIGURE 7.7 Tracking of an individual $n=1$ TAE resonance in the absence of energetic particle drive in the limiter phase of a JET ohmic discharge. Left: The spectrogram of the directly digitised magnetic perturbations showing the zigzag pattern of the probing perturbation launched by the TAE antenna; Right: The synchronously detected signal[*].

7.2 ENERGETIC PARTICLE-DRIVEN AE WITH FREQUENCIES LOCKED TO PLASMA EQUILIBRIUM

We now consider experiments on plasmas in toroidal confinement devices with energetic ions produced by auxiliary heating systems, NBI and/or ICRH. Energetic particle-driven AEs are very often observed in plasmas with energetic ions. Instabilities of AEs can be excited by energetic ions, such as fusion-generated alpha particles or produced by NBI and/or ICRH, if the following conditions are fulfilled:

a. the drift frequency due to the radial gradient of the energetic ion pressure (which constitutes the free energy source for such instabilities) is larger than AE eigenfrequency:

$$\omega_{*\alpha} \equiv -\frac{m}{r}\frac{cT_\alpha}{e_\alpha B_0}\frac{d\ln p_\alpha}{dr} > 2\pi f_{AE}, \tag{7.2}$$

b. the free energy of energetic ions can be released into the AE via wave-particle resonances, for example, via the Landau resonance between the AE and passing ions,

$$V_{\|\alpha} = V_A, \tag{7.3}$$

c. power transfer P_α from energetic ions to AE overcomes the AE damping by thermal plasma:

$$\gamma_\alpha = \frac{P_\alpha}{2W_{AE}} > |\gamma_{damp}|. \tag{7.4}$$

Condition (7.2) implies that, for a given alpha particle pressure profile, the particle-to-wave power transfer is positive if the wave has a high poloidal mode number m and/or the wave frequency f_{AE} is low enough. Condition (7.3) is easily satisfied for alpha particles in burning plasma of ITER and could be achieved in present-day machines with low magnetic fields and low V_A, for example, on spherical tokamaks. Finally, condition (7.4) selects Alfvén eigenmodes with the lowest $|\gamma_{damp}|$, which have the smallest threshold in the pressure gradient of energetic particles and are the easiest to excite. Note, however, that modes easiest to excite are not necessarily the ones providing the

---

[*] Reproduced from [A. Fasoli et al., *Plasma Phys. Control. Fusion* 44 (2002) B159], with the permission of IOP Publishing.

most significant transport of energetic particles in the non-linear phase. In scenarios with a positive magnetic shear, the weakest damping is exhibited by TAEs and EAEs.

In typical JET discharges, AEs are most often excited by ICRH-accelerated energetic hydrogen minority ions in the MeV energy range, while deuterium beams of energy 80–140 keV are sub-Alfvénic and provide damping for TAEs via the $V_{beam} = V_A/3$ resonance. Figure 7.8 shows a typical JET discharge with both NBI and ICRH heating. In this discharge, we see electromagnetic turbulence and MHD modes with rather significant amplitudes in the low-frequency range of ≤100 kHz. The zigzag trace seen in the frequency range of ~150–250 kHz shows the perturbation launched by the external TAE antenna probing the damping of stable TAEs. Unstable TAEs are excited by ICRH-accelerated H-minority ions and are seen in the frequency range of ~250–400 kHz as nine spectral lines parallel to each other, frequencies of which are slowly decreasing in time as the plasma mass density increases fuelled by NBI. The Doppler shift due to toroidal plasma rotation separates TAEs with neighbouring ns plasma by the local frequency of plasma rotation (see Appendix D). As TAE frequencies vary in time slowly due to the increase in plasma density, the modes belong to the FL non-linear scenario.

The most important properties of TAEs described in Chapter 6 could be validated from the FL scenarios. To begin, one can consider whether the frequency ranges of TAEs and EAEs observed experimentally are close to the ones predicted theoretically. For such comparison, we take JET discharge without significant toroidal plasma rotation so that the Doppler shift due to toroidal rotation is small and AE frequencies in the lab reference frame are close to those in the plasma reference frame. Figure 7.9 shows such comparison in one of the JET discharges with ICRH-accelerated ions [7.5].

As shown in Figure 7.9, both branches of weakly damped modes, TAEs and EAEs, are excited in JET discharge #40399 with elevated q-profile, $q_{min} \sim 1.7$. Both classes of modes closely follow the Alfvén scaling in time. Toroidal mode numbers of TAEs and EAEs are low and mostly positive, $n = 1, \ldots, 6$, but some modes had negative toroidal mode numbers at the beginning of the discharge. This effect was identified to be caused by the starting ICRH resonance position off-axis at the low field side so the pressure gradient of energetic ions was hollow. Figure 7.9 shows that the MHD modelling of the Alfvén continua for low toroidal mode numbers provides the characteristic frequency range of TAEs and EAEs in a satisfactory agreement with the experiment.

Next, the threshold for TAE excitation could be validated in a discharge with gradually increasing ICRH power that generates energetic trapped ions driving TAEs. In this case, it is instructive

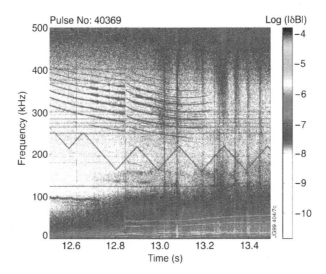

FIGURE 7.8 Magnetic spectrogram of ICRH-driven TAEs with different toroidal mode numbers seen in the frequency range of ~250–400 kHz in JET discharge #40369.

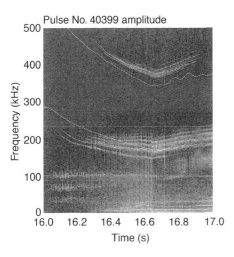
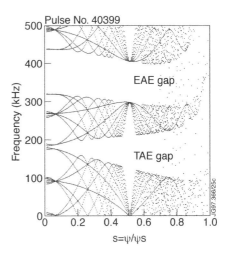

FIGURE 7.9 Left: Magnetic spectrogram showing TAEs and EAEs excited by ICRH energetic ions in JET #40399. The white lines show the computed centre-of-gap frequencies for TAE and EAE. Right: The computed Alfvén continuous spectra for $n = 1 \ldots 6$ as functions of radial variable $s \equiv \sqrt{\psi_{pol}/\psi_{pol}(a)}$, with TAE and EAE gaps clearly visible.

to consider a discharge with high NBI power applied, so that a large Doppler shift exists due to the NBI-driven toroidal rotation. Due to the Doppler shift caused by the plasma rotation, TAEs of different n's have their frequencies separated very well. Figures 7.10 and 7.11 show such JET case. Figure 7.10 displays the power wave-forms of ICRH, which excites TAEs, and of NBI, which provides TAE damping. The Doppler shifts make it easy to observe in Figure 7.11 which TAEs are excited first at the gradually increasing ICRH power, that is, which modes have the lowest excitation threshold in the fast particle pressure gradient.

As ICRH power increases, the following sequence is excited in TAEs with different toroidal mode numbers n: $n = 8$ at $t \approx 12.57$ s, then, after a very short delay, TAEs with $n = 9$ and $n = 7$ get excited, and after yet another short delay $- n = 10$ and $n = 6$. All these TAEs were excited when ICRH power was $P_{ICRH} \approx 4$–5 MW. Only after much longer delay of ≈ 50 ms, when ICRH power approached $P_{ICRH} \approx 6.5$ MW, TAEs with $n = 11$ and $n = 5$ were excited. After yet longer delay of ≈ 80 ms when ICRH power became steady-state at $P_{ICRH} \approx 7$ MW, TAE with $n = 4$ was excited. Because neither $n = 3$, from the low end, nor $n = 12$, from the high end, were driven unstable, the power $P_{ICRH} \approx 7$ MW could be considered as insufficient for exciting TAEs with $n = 3$ and $n = 12$. The unstable TAEs appearing in a sequence of certain toroidal mode numbers is consistent with the qualitative description of TAE drive in Chapter 6, which tells us that most effectively driven TAEs are those with their radial width equal to the width of fast ion drift orbit. The TAEs with different n's observed have their radial widths estimated as:

$$\Delta_{TAE} \sim \frac{r_{TAE}}{nq} \tag{7.5}$$

where the safety factor at the position of the $n = 8$ TAE is associated with $q = 8.5/8 \approx 1.06$ because this JET plasma had sawteeth and a $q = 1$ magnetic surface. Taking into account that this TAE magnetic surface is localised just outside the sawtooth inversion radius, that is, at $r_{TAE} \approx 30$ cm, the radial width of the $n = 8$ TAE can be estimated as $\Delta_{TAE} \approx 3.5$ cm. This width also provides an estimate of the drift orbit of ICRH-accelerated H-minority ions driving the TAEs.

By noting that the time axis in Figure 7.11 corresponds to the increase in β_{ICRH}, while the frequency axis due to the Doppler shift provides TAEs with increasing n's, we obtain a TAE instability zone displayed in Figure 7.11. This figure shows which critical thresholds are in β_{ICRH} for TAE with every n involved for this type of energetic particle population and in this type of discharge.

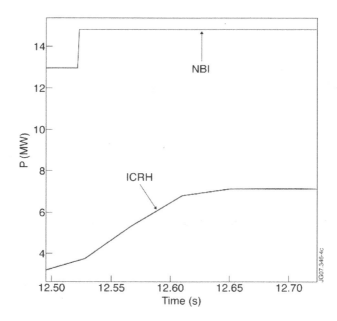

FIGURE 7.10 Temporal evolution of ICRH power accelerating H-minority ions and driving TAEs, and sub-Alfvénic NBI that provides a strong damping for TAEs. Figure 7.11 Magnetic spectrogram of ICRH-driven Alfvén instabilities in JET discharge #40329 with gradually increasing ICRH power.

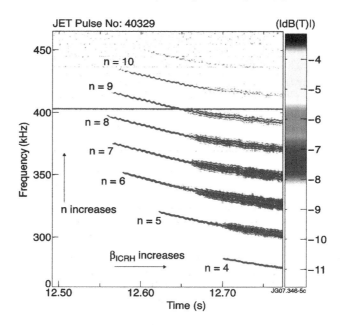

FIGURE 7.11 Magnetic spectrogram of ICRH-driven Alfvén instabilities in JET discharge #40329 with gradually increasing ICRH power.

Figure 7.11 shows that each spectral line of TAEs excited first saturates in amplitude and represents a narrow single eigenfrequency. However, as ICRH power increases further, each spectral line gets some side-bands with fine frequency splitting, and the TAE looks like a frequency band surrounding the eigenfrequency. For explaining such phenomenon, we have to consider the non-linear theory of wave-particle interaction. This will be the subject of Chapter 8.

Finally, we note an important possibility for plasma diagnostics based on experimental observations of AEs with FL evolution. Because FL scenarios could be modelled with linear spectral MHD codes computing the AEs supported by plasma equilibrium, it is relatively easy to develop MHD spectroscopy [7.13] via AEs observed experimentally [7.14,7.15]. This implies solving the inverse problem of obtaining information on plasma equilibrium from observed spectrum of AEs. The MHD spectroscopy via TAEs was developed first in the 1990s. Later, it became an important part of advanced tokamak developments aiming at obtaining internal transport barriers in tokamaks. The so-called Alfvén cascade (AC) eigenmodes [7.14,7.16], also called reversed-shear Alfvén eigenmodes, RSAEs, were employed in the MHD spectroscopy of reversed-shear plasma equilibria. Similar to TAEs, ACs belong to the FL class of Alfvén instabilities. In contrast to TAE, ACs are localised at the minimum of the safety factor, q_{min}, and frequency of an AC locks to the time-dependent $q_{min}(t)$ as $\omega_{AC} \approx |n - m/q_{min}(t)| \cdot V_A/R + \Delta\omega$, thus tracking the time evolution of q_{min}, see Chapters 9 and 10 for details.

7.3   FREQUENCY-SWEEPING ENERGETIC PARTICLE-DRIVEN AE

The history of FS instabilities driven by energetic ions started from an oscillatory "fishbone" instability, with the dominant mode numbers $n = 1$ and $m = 1$, first observed in experiments with perpendicular NBI on the Poloidal Divertor Experiment (PDX) tokamak [7.17]. The instability occurs in repetitive bursts, with the mode frequency decreasing by about a factor of 2 during each burst. Large fishbone bursts cause losses of NBI-produced energetic ions, thus reducing the efficiency of plasma heating. Experimentally, the fishbone mode structure was found to be of the "top hat" type similar to the internal kink mode [7.18] associated with the surface $q = 1$ in tokamak plasma. The frequency of the fishbone oscillations on PDX was found to be close to the magnetic precession frequency of the trapped beam ions, as well as to the diamagnetic frequency of thermal ions. Two different regimes have been identified for the linear phase of fishbone instability. The first regime of the so-called "precessional" fishbones [7.19] refers to the case when the mode frequency is much greater than the thermal ion diamagnetic frequency. In this case, trapped energetic ions via the resonance with their precessional motion destabilise the $n = 1$, $m = 1$ mode emerging from the Alfvén continuum. The mode frequency is essentially determined by the energetic particle population in this case, and the fishbone represents an energetic particle mode. In this case, the mode structure has singularities at the radii of local Alfvén resonances (5.23) and significant continuum damping. Such fishbones are excited at relatively high values of the energetic ion pressure which overcomes the threshold value as determined by the continuum damping. In the non-linear phase, such fishbones could sweep in frequency due to the re-distribution of energetic ions and the MHD nonlinearity of the continuum damping [7.20]. The second regime [7.21,7.22] corresponds to the case of comparable precessional frequency of the beam ions and the diamagnetic frequency of thermal ions. The mode frequency could then reside in a low-frequency "gap" in the continuum associated with the diamagnetic frequency, thus avoiding continuum damping.

Fishbone oscillations were later observed in many other tokamaks with significant populations of energetic ions produced by ICRH, perpendicular and parallel NBI, see review [7.23] and references therein. Fishbones driven by supra-thermal electrons resulting from ECRH and LHCD were also observed and explained [7.24]. Significant further advances were achieved in theory and modelling of the fishbones, see [7.25,7.26] and references therein. We note, however, that in future large machines with significant populations of fusion-generated alpha particles, fishbones will be driven by alpha particles with characteristic energies of ~400 keV [7.27]. Because this energy range is well below the birth energy of 3.52 MeV, the possible re-distribution or losses of energetic resonant ions caused by the fishbone instability are not as dangerous as TAEs or other AEs with frequencies much higher than the fishbone frequency and evolving alpha particles with energy ~1 MeV and more.

Meanwhile, another type of energetic ion-driven instability with sweeping frequency, but of much higher frequency, was reported from DIII-D experiment with NBI [7.28]. This instability on DIII-D started from a frequency of ~90 kHz, which is much higher than the fishbone frequency of ~20 kHz, and very close to the TAE frequency of ~110 kHz, and swept down in frequency by a factor of two in a very short time of ~2.5 ms. Due to the very fast sweep down in frequency, these beam-driven modes were called "chirping" modes. Toroidal mode numbers of the chirping modes were positive and in the range of $n = 1, ..., 8$. The beam parallel velocity in this DIII-D experiment was sub-Alfvénic, $V_{beam} \approx 0.3 - 0.5\ V_A$, but beam pressure was large, $\langle \beta_{beam} \rangle \geq 1\%$. Moreover, the toroidal plasma rotation was also large. The chirping modes transported beam ions to the edge of the plasma. Bursts of chirping modes correlated with drops in neutron emission by 10%, which was much higher than that of the fishbones, $\leq 2\%$. Because these plasmas had significant beam-plasma neutrons, the reductions in neutron yield indicated the loss of beam from the plasma centre much higher than that during the fishbone.

The instability of chirping modes was then observed on JT-60U [7.6,7.29] and on small spherical tokamak START [7.30]. Despite the difference in the plasma size, both machines used NBI with velocities close to, or exceeding, the Alfvén velocity. The experiment on JT-60U used negative NBI of very high energy of ~360 keV at magnetic field of ~1.2 T, while the experiment on START used very low magnetic field in the range of ~0.2–0.4 T and H beam with energy of ~30 keV.

Two large spherical tokamaks, NSTX [7.31] and MAST [7.32], were built, and the chirping modes were found to become a dominant type of beam-driven instabilities in STs. Figure 7.12 illustrates the various Alfvén instabilities that could exist in a typical MAST discharge. The discharge in Figure 7.12 was in L-mode phase, with the following parameters: $I_p \approx 730$ kA, $B_T (0) = 0.5$ T, $\langle n_e \rangle \approx 1.6 \times 10^{19}$ m^{-3}, $\beta \sim 2.2\%$, and $P_{NBI} \approx 1.7$ MW.

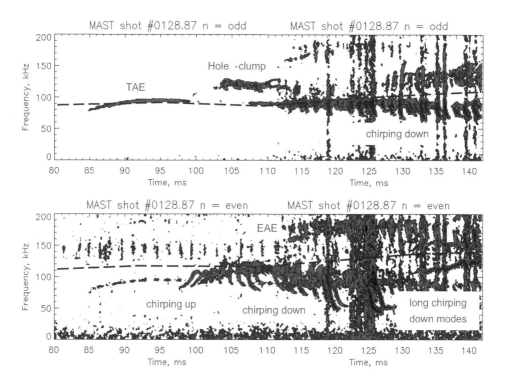

FIGURE 7.12 Magnetic spectrograms for $n =$ odd (top) and $n =$ even (bottom) components of the outer mid-plane Mirnov coil signal for MAST discharge #12887. Different types of Alfvénic modes are seen. The dashed line represents time evolution of the TAE-gap centre with Doppler shift taken into account.

Figure 7.12 presents magnetic spectrograms for n=odd and n=even components of Mirnov coil signal for MAST discharge #12887 showing examples of various beam-driven modes. In this discharge, NBI was applied at 80 ms exciting TAE in an FL regime first. This mode was replaced by a set of n = odd bursts of up-down symmetric "clump-hole" modes (to be explained in next chapter) and n = even modes that first sweep up, and then sweep down in frequency. Later in the discharge, the chirping-down modes are seen with a long chirping-down phase in the n = even component.

Apart from the super-Alfvénic beam, the beam pressure was rather high on MAST, similar to [7.28]. Figure 7.13 shows values of β_{beam} and $\beta_{\text{therm}} = \beta_e + \beta_i$ in a set of MAST discharges with NBI, which were calculated with TRANSP and EFIT. Each point represents one TRANSP/EFIT output and the lines represent transport trajectories for some individual discharges. The dashed line shows the margin of $\beta_{\text{beam}}/\beta_i \approx \dfrac{2\beta_{\text{beam}}}{\beta_{\text{therm}}} = 1$. Although beam fraction was as high as ~80% in some cases, these high values were not typical for MAST.

Figure 7.14 provides a larger picture of long chirping-down modes. It is seen that the initial FS is nearly linear in time, and it gives the frequency deviation from the starting point by a factor of 2 (e.g., from ~120 kHz down to ~60 kHz) in ~1 ms. The mode amplitude increases significantly at the end of the linear chirping phase. Then, the sweeping rate decreases significantly, from ~60 kHz down to ~40 kHz in ~2 ms, while the amplitude does not change much. Explanation of this long chirping-down modes is not an easy task, and the possible options will be described in Chapter 8.

Figure 7.14 Magnetic spectrogram showing long chirping modes with n=even component in MAST discharge #12878.

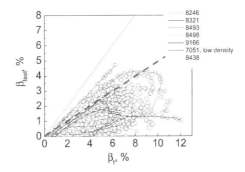

FIGURE 7.13 Typical values of β_{beam} and β_{therm} in MAST discharges with NBI (calculated with TRANSP and EFIT). Each point represents one TRANSP/EFIT output and the lines represent trajectories for individual discharges. The dashed line shows the $\beta_{\text{beam}}/\beta_i = 1$ margin.

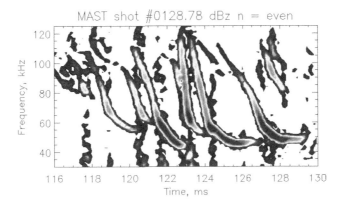

FIGURE 7.14 Magnetic spectrogram showing long chirping modes with n=even component in MAST discharge #12878.

In recent years, the phenomenon of rapid chirping of energetic beam-driven modes has been found on stellarators [7.33,7.34], in a dipole experiment [7.35], and in JET experiments with ICRH-accelerated ions [7.36]. Furthermore, in addition to the fishbones excited by fast electrons produced with ECRH and LHCD [7.24], rapid FS instabilities driven by runaway electrons were observed in DIII-D [7.37], as well as in mirror-confined plasma sustained by high-power microwaves [7.38]. The widespread of the FS phenomena suggests that all of them could be possible to explain within a framework of a generic theoretical model.

REFERENCES

1. D.W. Ross et al., *Phys. Fluids* **25** (1982) 652.
2. K. Appert et al., *Plasma Phys.* **24** (1982) 1147.
3. A. Fasoli et al., *Nucl. Fusion* **35** (1995) 1485.
4. S.E. Sharapov et al., *Nucl. Fusion* **53** (2013) 104022.
5. W. Kerner et al., *Nucl. Fusion* **38** (1998) 1315.
6. K. Shinohara et al., *Nucl. Fusion* **41** (2001) 603.
7. A. Fasoli et al., *Phys. Rev. Lett.* **75** (1995) 645.
8. A. Fasoli et al., *Phys. Plasmas* **7** (2000) 1816.
9. A. Fasoli et al., *Phys. Rev. Lett.* **76** (1996) 1067.
10. P. Puglia et al., *Nucl. Fusion* **56** (2016) 112020.
11. A. Fasoli et al., *Plasma Phys. Control. Fusion* **52** (2010) 075015.
12. A. Fasoli et al., *Nucl. Fusion* **36** (1996) 258.
13. J.P. Goedbloed et al., *Plasma Phys. Control. Fusion* **35** (1994) B277.
14. S.E. Sharapov et al., *Phys. Lett.* **A289** (2001) 127.
15. A. Fasoli et al., *Plasma Phys. Control. Fusion* **44** (2002) B159.
16. S.E. Sharapov et al., *Phys. Plasmas* **9** (2002) 2027.
17. K. McGuire et al., *Phys. Rev. Lett.* **50** (1983) 891.
18. M.N. Bussac et al., *Phys. Rev. Lett.* **35** (1975) 1638.
19. L. Chen, R.B. White, and M.N. Rosenbluth, *Phys. Rev. Lett.* **52** (1984) 1122.
20. A. Odblom et al., *Phys. Plasmas* **9** (2002) 155.
21. B. Coppi and F. Porcelli, *Phys. Rev. Lett.* **57** (1986) 2272.
22. B. Coppi, S. Migliuolo, and F. Porcelli, *Phys. Fluids* **31** (1988) 1630.
23. W.W. Heidbrink and G. Sadler, *Nucl. Fusion* **34** (1994) 535.
24. F. Zonca et al., *Nucl. Fusion* **47** (2007) 1588.
25. L. Chen and F. Zonca, *Nucl. Fusion* **47** (2007) S727.
26. B.N. Breizman and S.E. Sharapov *Plasma Phys. Control. Fusion* **53** (2011) 054001
27. B. Coppi and F. Porcelli, *Fusion Technology* **13** (1988) 447.
28. W.W. Heidbrink, *Plasma Phys. Conuol. Fusion* **37** (1995) 937.
29. Y. Kusama et al., *Nucl. Fusion* **39** (1999) 1837.
30. M.P. Gryaznevich and S.E. Sharapov, *Nucl. Fusion* **40** (2000) 907.
31. E.D. Fredrickson et al., *Nucl. Fusion* **46** (2006) S926.
32. M.P. Gryaznevich, S.E. Sharapov, *Nucl. Fusion* **46** (2006) S942.
33. K. Toi et al., *Plasma Phys. Control. Fusion* **53** (2011) 024008.
34. A.V. Melnikov et al., *Nucl. Fusion* **58** (2018) 082019.
35. D. Maslovsky et al., *Phys. Plasmas* **10** (2003) 1549.
36. H.L. Berk et al., *Nucl. Fusion* **46** (2006) S888.
37. A. Lvovskiy et al., *Nucl. Fusion* **59** (2019) 124004.
38. A.G. Shalashov et al., *Plasma Phys. Control. Fusion* **61** (2019) 085020.

8 Non-linear Evolution of Coupled Energetic Particle Populations and Energetic Particle-Driven Modes

8.1 GENERIC BERK-BREIZMAN THEORY ON THE NEAR-THRESHOLD WAVE EXCITATION BY ENERGETIC PARTICLES

In a typical plasma heating scenario with energetic ions, a gradual build up of energetic ion pressure occurs, so that the energetic ion drive of an AE, γ_L, increases in time at unchanged AE damping, γ_d. At the AE linear instability threshold, an exact balance between the AE drive and damping is achieved,

$$\gamma_L = |\gamma_d|. \tag{8.1}$$

As soon as the drive overcomes the instability threshold (8.1), the AE gets excited with the net linear growth rate determined by the difference between γ_L and γ_d. In a theory developed in Ref. [8.1], excitation scenarios and early non-linear evolution of any energetic particle-driven modes are of a universal type in the near-threshold regime:

$$\gamma \equiv \gamma_L - |\gamma_d| \ll |\gamma_d| < \gamma_L. \tag{8.2}$$

In this case, the small net growth rate becomes comparable to the effective collisional frequency restoring the energetic particle distribution,

$$|\gamma_L - \gamma_d| \approx \nu_{\text{eff}} \tag{8.3}$$

and the electric field of the perturbation that tends to flatten the distribution function at the resonance competes with the source of the energetic particles that replenishes the unstable distribution function.

The Berk–Breizman near-threshold non-linear theory [8.1] was developed for one-dimensional (1D) bump-on-tail (BOT) instability of a single electrostatic mode when this mode gets excited via Landau resonance due to the positive gradient, $\dfrac{\mathrm{d}F}{\mathrm{d}v} > 0$, at the resonance (as Figure 8.1 shows),

$$v = \omega / k, \tag{8.4}$$

between the wave and fast electron population. The mode damping is an essential part of the theory, and the damping rate γ_d was kept constant in Ref. [8.1] throughout the mode evolution.

The non-linear scenarios [8.1] were found in many various types of wave-particle resonance systems with an excitation threshold, making it clear that the Berk–Breizman theory is very generic. Here, we present the key points of the Berk–Breizman near-threshold theory, and illustrate its validity via some experiment-to-theory comparison examples.

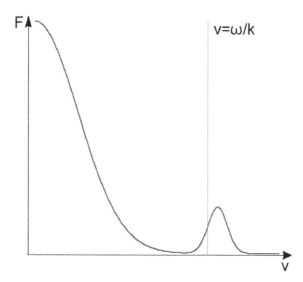

FIGURE 8.1 Schematic illustration of a distribution function of with the bump-on-tail in the high-energy range close to the wave resonance velocity ω/k.

In order to accommodate the collisions, a 1D Fokker–Planck equation is used for describing energetic particle distribution $F(x, v, t)$ coupled to the electric field of the mode

$$E = \frac{1}{2}\left[\hat{E}(t)e^{i(kx-\omega t)} + c.c \right] \tag{8.5}$$

The Fokker–Planck equation to be solved has the form

$$\frac{\partial F}{\partial t} + v\frac{\partial F}{\partial x} + \frac{e}{2m}\left[\hat{E}(t)e^{i(kx-\omega t)} + c.c \right]\frac{\partial F}{\partial v} = \frac{dF}{dt}\Big|_{\text{coll}} \tag{8.6}$$

together with Maxwell's equations for electric field ($\delta B = 0$ for this problem):

$$\left[-i\omega_{\text{pe}} \frac{\partial \hat{E}(t)}{\partial t}e^{i(kx-\omega t)} + c.c \right] + 4\pi \frac{\partial j_f}{\partial x} = 0, \tag{8.7}$$

where j_f is the energetic particle contribution to perturbed current produced by the wave. Only resonant particles contribute to the wave evolution, so the relevant part of the collisional operator can be taken in the vicinity of the resonance as

$$\frac{dF}{dt}\Big|_{\text{coll}} = \alpha^2\left(\frac{\partial F}{\partial u} - \frac{\partial F_0}{\partial u} \right) + v^3\left(\frac{\partial^2 F}{\partial u^2} - \frac{\partial^2 F_0}{\partial u^2} \right) - \beta(F - F_0) \tag{8.8}$$

where

$$u = kv - \omega, \tag{8.9}$$

and α, v, β are coefficients of drag, diffusion, and the Krook operator, respectively. For the Krook operator, the coefficient is constant, but for drag and diffusion these are taken at the resonant point.

The equation with both the perturbed electric field of the wave and the collisions is:

$$\frac{\partial F}{\partial t} + \left(\frac{u+\omega}{k} \right)\frac{\partial F}{\partial x} + \frac{ek}{2m}\left[\hat{E}(t)e^{i(kx-\omega t)} + c.c \right]\frac{\partial F}{\partial u} - v^3\frac{\partial^2 F}{\partial u^2} - \alpha^2\frac{\partial F}{\partial u}$$

$$+\beta F = -v^3\frac{\partial^2 F_0}{\partial u^2} - \alpha^2\frac{\partial F_0}{\partial u} + \beta F_0 \tag{8.10}$$

We represent the distribution function as a Fourier series

$$F = F_0 + f_0 + \sum_{n=1}^{\infty} \left[f_n \exp(in\psi) + c.c \right]$$

(8.11)

$$\psi = kx - \omega t$$

so that the wave equation relating the field and the fast particle current becomes

$$\frac{\partial \hat{E}}{\partial t} + 4\pi e \frac{\omega}{k^2} \int f_1 du + \gamma_d \hat{E} = 0$$

(8.12)

Consider time scales shorter than non-linear bounce period of the wave. With the distribution function being not too significantly perturbed, that is, within the ordering

$$F_0 \gg f_1 \gg f_0, f_2$$

(8.13)

where f_1 admits a power series in $\hat{E}(t)$

$$f_1 \approx C_1 \hat{E} + C_3 \hat{E}^3 + \dots$$

(8.14)

which allows the first-order (cubic) nonlinearity to be captured by the following truncated Fourier expansion:

$$\frac{\partial f_0}{\partial t} - v^3 \frac{\partial^2 f_0}{\partial u^2} - \alpha^2 \frac{\partial f_0}{\partial u} + \beta f_0 = -\frac{ek}{2m} \left(\hat{E} \frac{\partial f_1^*}{\partial u} + c.c \right)$$

$$\frac{\partial f_1}{\partial t} + iuf_1 - v^3 \frac{\partial^2 f_1}{\partial u^2} - \alpha^2 \frac{\partial f_1}{\partial u} + \beta f_1 = -\frac{ek}{2m} \hat{E} \frac{\partial}{\partial u} (F_0 + f_0 + f_2)$$

(8.15)

$$\frac{\partial f_2}{\partial t} + 2iuf_2 - v^3 \frac{\partial^2 f_2}{\partial u^2} - \alpha^2 \frac{\partial f_2}{\partial u} + \beta f_2 = -\frac{ek}{2m} \hat{E}^* \frac{\partial f_1}{\partial u} + O(\hat{E}f_3)$$

The wave amplitude equation near the threshold is obtained then assuming the smallness of the net growth rate (8.2), and it takes the form of cubic non-linear integral equation

$$\frac{dA}{d\tau} = A(\tau) - \frac{1}{2} \int_0^{\tau/2} dz z^2 A(\tau - z) \int_0^{\tau - 2z} dx \, e^{-\hat{v}^3 z^2 (2z/3 + x) - \hat{\beta}(2z + x) + i\hat{\alpha}^2 z(z + x)} \times A(\tau - z - x) A^*(\tau - 2z - x)$$

(8.16)

where

$$A = \left[ek\hat{E}(t) / m(\gamma_l - \gamma_d)^2 \right] \left[\gamma_l / (\gamma_l - \gamma_d) \right]^{1/2}, \tau = (\gamma_l - \gamma_d)t, \hat{v}^3 = v^3 / (\gamma_l - \gamma_d)^3,$$

$$\hat{\alpha} = \alpha / (\gamma_l - \gamma_d)^2, \hat{\beta} = \beta / (\gamma_l - \gamma_d) \text{ and } \gamma_l = 2\pi^2 (e^2 \omega / mk^2) \partial F_0 (\omega/k) / \partial v.$$

Here, the time integration corresponds to the "memory" of the system of coupled energetic particles and the wave because the temporal mode evolution at some time slice depends on the previous mode evolution integrated over the entire time of the instability. In the absence of the drag, $\alpha = 0$, the exponent in (8.16) represents an effective window back in time, during which the information on the history of the mode evolution is essential. The argument of the "history exponent" is determined by v and β, and it determines how far back in time the history matters for the solution.

For different values and the three different types of the effective collisionality, (8.16) exhibits both "soft" and "hard" early non-linear scenarios of excitation. In the "soft" scenario, the evolution of the unstable system tends to return to the original state by flattening the distribution function in the resonance area. The mode amplitude evolves to a low level reflecting the closeness to the threshold.

In the "hard" excitation scenario, the instability pushes the system further away from the original state, and the mode amplitude "explodes" in a finite time. In this case, the low-order cubic nonlinearity becomes insufficient soon after the mode excitation, and one needs a fully non-linear model for describing the long-term mode evolution. We now consider non-linear solutions of (8.16) in some important cases and illustrate theory-to-experiment comparison.

8.1.1 Non-linear Scenarios of the Mode Evolution Described by (8.16)

One of the most important cases to consider in numerical modelling of (8.16) is when an effective diffusion replenishes the unstable distribution of energetic particles. In this case, (8.16) has no drag or Krook effects, $\alpha = \beta = 0$, and the non-linear solutions are only determined by the diffusion term that depends on ν. Figure 8.2 shows various solutions of (8.16) for the mode amplitude as the effective diffusion collisionality ν changes. The amplitude solutions shown are the envelopes of the high-frequency waves, with the characteristic times determined by the inverse growth rate of the mode.

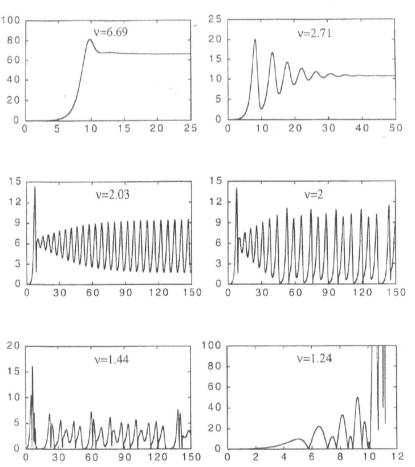

FIGURE 8.2 Numerical solutions for $|A|$ of Eq. (8.16) with diffusion only for different values of effective normalised collisionality ν. Here, $t \equiv \gamma t$.

We can see from Figure 8.2 that for high values of ν, $\nu=6.69$ and $\nu=2.71$, non-linear solutions are established with a steady-state mode amplitude after the initial linear exponential growth and some transient oscillatory phase. The steady-state solutions are expected as there is a "minus" sign in front of the non-linear term, which tends to compensate the linear growth rate term (first in the right-hand-side of (8.16)). The "history exponent" provides a very narrow time window at high ν, so the time integration should not cause surprises.

For lower values of ν, $\nu=2.03$ and $\nu=2$, the non-linear solution exhibits amplitude modulation [8.2]. Such a scenario is much less obvious than the steady-state one. The oscillatory character of the mode evolution means that the non-linear term in (8.16) changes its sign quasi-periodically with time. This becomes possible because the "history exponent" at lower ν significantly expands the time window, and the product of three complex amplitudes taken at different times, which can generate both positive and negative values, becomes essential in the time integral.

We can also see from Figure 8.2 that at $\nu=2$ the amplitude modulation becomes very deep, that is, comparable to the amplitude itself. This causes the non-linear system to go through the zero amplitudes because of two different mechanisms: the high-frequency solution has zeroes itself, and the envelope in the form of the amplitude modulation provides another set of zeroes. Under these conditions, the non-linear system mixes the phase information around the zero values of the amplitude and picks up the phase randomly, as shown in Figure 8.3. This results in a "chaotic" mode evolution [8.3] that starts at $\nu \leq 2$ and becomes very well-determined at $\nu=1.44$ (see Figure 8.2).

For yet lower value of the diffusion coefficient, $\nu=1.24$, the solution becomes explosive, that is, the mode amplitude becomes infinite during finite time, as shown in Figure 8.2. This means that the theory limited by the low-order cubic nonlinearity (8.16) becomes insufficient, and fully non-linear approach should be employed for describing the wave-particle coupled system. It was found in Ref. [8.4] that a fully non-linear approach gives, beyond the explosive scenario, a spontaneous creation of frequency-sweeping holes and clumps in the energetic particle distribution (to be discussed later).

In summary, four essentially different non-linear scenarios were found for (8.16) controlled by the parameter of the diffusive collisionality ν:

a. steady-state;
b. periodically modulated;
c. chaotic;
d. explosive.

Another important case to consider in numerical modelling of (8.16) is when drag replenishes the unstable distribution of energetic particles. In this case, Eq. (8.16) has no diffusion or Krook effects, $\nu=\beta=0$, and the non-linear solutions are only determined by the drag term that depends on α.

Only explosive solutions of (8.16) were found in this case [8.5] in contrast to the diffusion. The inclusion of a drag term introduces an oscillatory dependence to the integral in (8.16), instead of

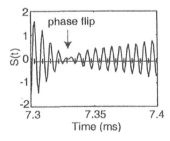

FIGURE 8.3 Zoom of numerical solution of (8.16) for wave with a deep amplitude modulation. The phase flip is seen leading to a random phase pick-up at the maximum of the modulation causing chaotic mode evolution on a longer time scale.

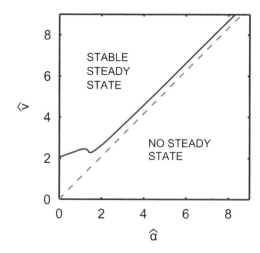

FIGURE 8.4 Displays the boundaries in parameter space that give stable, unstable, and no steady-state solutions to (8.16) with drag and diffusion. The unstable solution lies in between the solid and dashed lines [8.5].

the exponent in the case of diffusion. The oscillatory dependence has a profound effect on the non-linear evolution of the mode amplitude as the integral in (8.16) can easily change sign. The contributions to the integral from the time intervals with the sign "plus" exceed those with the sign "minus" so that the amplitude cannot saturate. Further description of the explosive scenario requires a fully non-linear approach.

In the presence of both drag and diffusion, Eq. (8.16) exhibits a variety of non-linear scenarios. Figure 8.4 shows the summary of non-linear scenarios found within the framework of (8.16) with the drag and diffusion effective collisionality terms.

8.1.2 Non-linear Scenarios Observed Experimentally for TAEs Excited with Energetic Tail Ions Resulted from the RF-Diffusion

TAEs excited by ICRH-accelerated ions is the best case for demonstrating the effect of a diffusive-type replenishment of the energetic ion distribution on the non-linear evolution of TAE. The energetic ion tail in ICRH case is built up by RF-diffusion given by Eq. (3.8), which has an effective frequency ν replenishing the distribution about ten times higher than that of Coulomb collisions [8.2]. All but one explosive non-linear scenarios of wave-particle interaction predicted by the Berk–Breizman near-threshold theory were validated experimentally in plasmas of JET.

In specific JET experiment [8.3], TAEs were excited by energetic H-minority tail ions accelerated by ICRH with gradually increasing ICRH power, as shown in Figure 8.5 (top). Due to the increase in ICRH power, the associated net growth rate of TAE makes the mode to go through the sequence of the non-linear saturated scenarios displayed in Figure 8.5 (bottom). First, we see a steady-state saturation of the TAE amplitude in the magnetic spectrogram. Then, from ~44.28 to ~44.4 s, the fine "pitchfork" splitting of the TAE spectral lines is observed. Finally, the modes go into the "chaotic" non-linear scenario, with very noisy spectral lines, widths of which remain the same as the "pitchfork" splitted TAE widths.

During the "chaotic" phase of the TAE excited, the phase analysis becomes complicated for measuring phase shift between the signals seen in toroidally separated Mirnov coils for determining toroidal mode numbers n. This phenomenon shown in Figure 8.6 results from the phase flips illustrated in Figure 8.3.

The different non-linear scenarios observed in Figure 8.5 follow one-by-one in time showing each of the four scenarios during relevant time intervals. Figure 8.7 shows the raw data from Mirnov coils corresponding to these time intervals in the magnetic spectrogram.

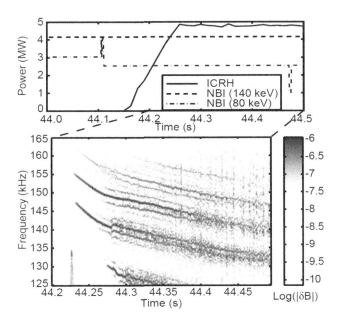

FIGURE 8.5 Top: the wave-forms of ICRH and NBI auxiliary heating power in JET discharge #49447. Bottom: Amplitude spectrogram of magnetic fluctuations showing the various non-linear scenarios of TAE evolution. The TAEs have toroidal mode numbers ranging from $n = 3$ to 6 [8.3].

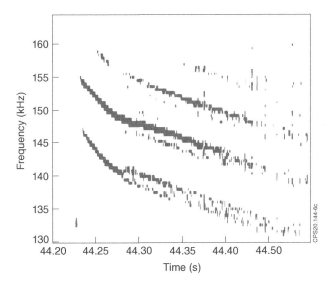

FIGURE 8.6 Phase magnetic spectrogram for determining toroidal mode numbers of TAEs in JET pulse #49447. The phase analysis becomes difficult during the chaotic TAE evolution.

A direct comparison between Figures 8.5 and 8.7 and the theoretical Figure 8.2 shows a remarkable similarity in the types of the non-linear evolution of the modes, one of which is driven by the BOT in the Berk–Breizman theory, and the other one is TAE driven by ICRH-accelerated ions in JET. Such a similarity could be explained by the structure of the wave-particle resonance for TAE and ICRH-accelerated ions. Because ICRH generates energetic trapped ions with their banana tips at the vertical line $\omega = \omega_{\mathrm{BH}}(R)$, all such ions have their pitch-angles $\Lambda = \mathrm{const}$ ($\Lambda = 1$ for on-axis ICRH). The wave-particle resonance at a fixed value of Λ takes the form of a 1D line in the (E,

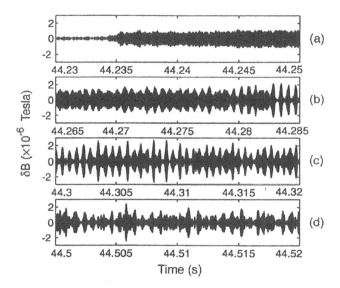

FIGURE 8.7 Raw data for JET pulse #49447 showing amplitude of TAE magnetic perturbations for different time intervals corresponding to TAE non-linear scenario of (a) steady-state type, (b) the amplitude modulation, (c) deep amplitude modulation, and (d) chaotic amplitude evolution.

P_φ)-space, and the physics of the resonant interaction is determined entirely by the motion of energetic particles across the resonance. This mechanism is similar to the 1D Vlasov BOT problem solved in the Berk–Breizman theory.

The explosive scenario was not achieved in the experiment shown in Figures 8.5–8.7. This is difficult to achieve with ICRH-driven TAEs dominated by the RF-diffusion at the resonance due to a strong RF-driven pitch angle scattering. However, the explosive scenario is much easier to obtain for TAEs driven by NBI-produced energetic ions when a dominant drag replenishment of the beam distribution is achieved.

8.2 SPONTANEOUS GENERATION OF HOLES AND CLUMPS IN ENERGETIC PARTICLE DISTRIBUTION BEYOND THE EXPLOSIVE SCENARIO AND FREQUENCY-SWEEPING MODES

It was noted above that the cubic non-linearity in (8.16) is insufficient for describing the explosive scenario. In a long-time fully non-linear modelling [8.4], long-living holes and clamps in the energetic ion distribution were observed for the first time. Then, to accommodate all types of the effective collisionality restoring the distribution function, a fully non-linear 1D model BOT was developed for the BOT instability including the effects of dynamical friction (drag) and velocity space diffusion on the energetic particles driving the wave. The BOT results show that in the early non-linear phase of the instability, the drag facilitates the explosive scenario of the wave evolution. Later, the drag effect leads to the creation of phase space holes and clumps moving away from the original eigenfrequency. The combined effect of drag and diffusion produces a diverse range of non-linear scenarios, including "hooked" frequency chirping and undulating regimes.

8.2.1 FULLY NON-LINEAR 1D BUMP-ON-TAIL (BOT) MODEL FOR LONG-TIME NON-LINEAR SCENARIOS BEYOND THE EXPLOSIVE PHASE

The fully non-linear system consists of a purely electrostatic wave in a plasma of three species. The first two are the thermal plasma ions and electrons and the third is a low density population of fast electrons that are subject to weak collisions (much less than the background species) and whose distribution function is treated kinetically.

In the normalised units of the Berk–Breizman theory, the starting set of coupled non-linear equations has the form [8.6]:

$$\frac{\partial F}{\partial t} + u\frac{\partial F}{\partial \xi} - \frac{1}{2}\left[\omega_B^2 e^{i\xi} + c.c.\right]\frac{\partial F}{\partial u} = \frac{dF}{dt}\bigg|_{coll} \tag{8.17}$$

$$\frac{\partial \omega_B^2}{\partial t} - \frac{4\pi|e|^2}{m_e}\frac{\omega_{pe}}{k}\int f_1\, du + \gamma_d\omega_B^2 = 0, \tag{8.18}$$

where the collisionality terms are given by (8.8), $\xi \equiv kx - \omega t$, and $\omega_B^2 = |e|\,kE_1/m_e$ is the non-linear bounce frequency of the electrons trapped in the field of the wave.

For describing holes and clumps identified in Ref. [8.4], we represent the frequency in the form $\omega \equiv \omega_{pe}\delta + \omega(t)$. As the holes and clumps evolve, the electric field becomes a sum of non-interacting BGK modes with time-dependent frequencies, so the envelope takes the form

$$E_n(t) = \sum_j \hat{E}_{n,j}(t)\exp\left[-in\int_{t_0}^t \delta\omega_j(t')dt'\right], \tag{8.19}$$

and the distribution function can be written in a similar manner.

To fully explore the non-linear system (Eq. 8.17 and 8.18), a numerical scheme similar to the one developed in Ref. [8.7] was employed for long-term evolution of the system. More specifically, the Fourier series representation of F in space transformed (8.17) into a set of coupled partial differential equations in t and u. By Fourier transforming in velocity, a set of advection equations was obtained for numerical processing with the BOT code. A fixed near-threshold parameter, $|\gamma_d|/\gamma_L = 0.9$ will be used for BOT simulations throughout this section, as Figure 8.8. shows a satisfactory agreement with more demanding case of $|\gamma_d|/\gamma_L = 0.99$.

Figure 8.9 shows the result of the BOT code for long-time evolution of the non-linear system (8.17) and (8.18) in the collision-less limit.

The holes and clumps move away from the original resonance, as shown schematically in Figure 8.10. This motion is almost adiabatic and preserves the value of the distribution function for

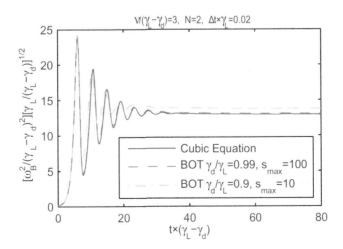

FIGURE 8.8 Comparison of the bump-on-tail simulation with the cubic equation (8.16) in the near-threshold regime with diffusion for two near-threshold parameters $|\gamma_d|/\gamma_L$.

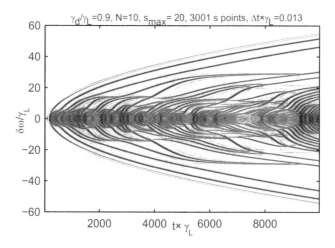

FIGURE 8.9   Spectrogram of the electric field amplitude E_1 for the collision-less case close to the threshold. The white line is the best $t^{1/2}$ fit passing through the upper and lower frequency-sweeping structures[*].

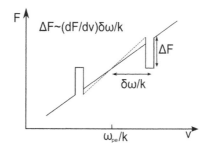

FIGURE 8.10   Cartoon illustrating the motion of holes and clumps and the wake dotted line that steepens the distribution function, creating a favourable environment for instability[*].

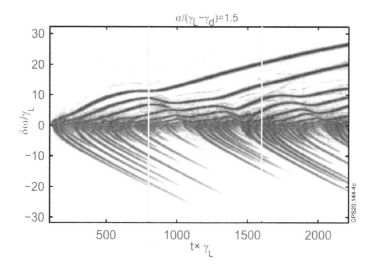

FIGURE 8.11   Spectrogram of the electric field amplitude showing chirping asymmetry for the pure drag case[*].

---

[*] Reproduced from [M.K. Lilley et al., *Phys. Plasma* 17 (2010) 092305], with the permission of AIP Publishing.

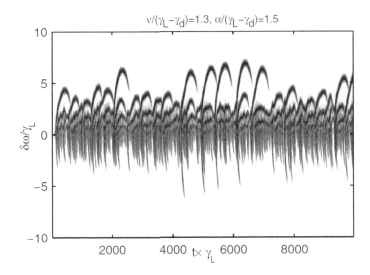

FIGURE 8.12 Spectrogram of the electric field amplitude showing the "hooked" frequency spectrum with drag and diffusion.

the particles trapped by the wave. We note that within this simplified theory there is no limit to the extent of the chirp, which is indeed supported by the simulation. The chirping behaviour shows the correct $t^{1/2}$ scaling [8.4].

The drag introduces a preferred direction of particle flow into the system (from high-energy source to low-energy sink). Consequently, one can expect the hole-clump symmetry observed in Figure 8.9 to be broken when drag is introduced, which is indeed the case as Figure 8.11 shows.

In the general case of both drag and diffusion affecting the distribution function at the resonance, more sophisticated patterns of the frequency sweeping are seen. One of this is the "hooked" frequency sweep shown in Figure 8.12.

8.2.2 NON-LINEAR FREQUENCY-SWEEPING SCENARIOS OBSERVED EXPERIMENTALLY FOR TAEs EXCITED WITH ENERGETIC IONS RESULTED FROM THE DRAG RELAXATION

Among all machines exhibiting the frequency-sweeping spectra, spherical tokamaks (START, MAST, NSTX) with super-Alfvénic NBI are the best test beds for investigating such phenomena, as discussed in Section 7.3.

The exceptional conditions on STs are determined by the low magnetic fields of STs and low values of T_e at the beginning of NBI, so that the following ordering is valid for the beam ions:

$$V_{crit} \ll V_{res} = V_A \le V_0, \tag{8.20}$$

where V_{crit} is the beam speed corresponding to the critical energy (3.3), and V_0 is the initial speed of the injected beam. Under the conditions (8.20), the beam passes through the principal resonance $v_{\parallel b} = V_A$ because of the beam slowing-down due to the electron drag. Figure 8.13 schematically shows that a short blip of NBI moves first uni-directionally from the injection high-energy range to low energy due to the drag, without changing its shape because the diffusion effect is small. Only at the low beam energy comparable to E_{crit}, the shape of the beam blip diffuses, and this diffusion is not uni-directional anymore. Because the resonance area E_{res} comes in this case to the region of the dominant beam relaxation due to the drag, we have a perfect drag scenario similar to that investigated in the 1D BOT model. Frequency-sweeping modes associated with the hole-clump generation are inevitable in such a scenario, which is what we observe in STs with super-Alfvénic NBI.

Figure 8.14 shows magnetic spectrogram from MAST experiment, in which super-Alfvénic NBI drives Alfvén instability when the resonance $V_{\|beam} = V_A$ is in phase space region dominated by electron drag of the beam ions. It is seen that FS modes dominate the spectrum, with some modes sweeping in frequency to a very long range of $|\delta\omega / \omega| \cong 0.5$. Modelling with the HAGIS code [8.8]

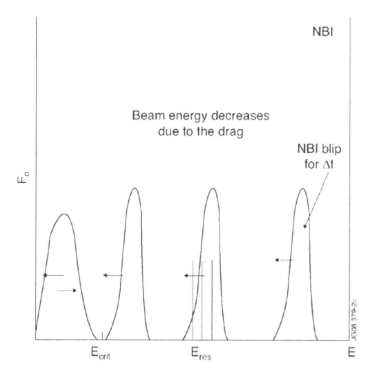

FIGURE 8.13   Schematic illustration of the beam blip evolution with time in the case of ordering (8.20) typical for STs.

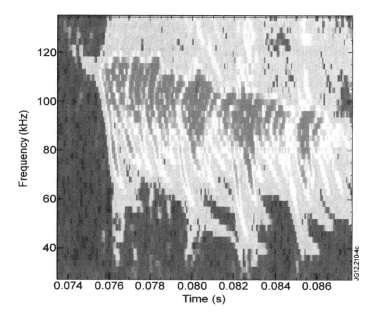

FIGURE 8.14   Spectrogram showing FS Alfvén modes driven by NBI in MAST discharge #27177.

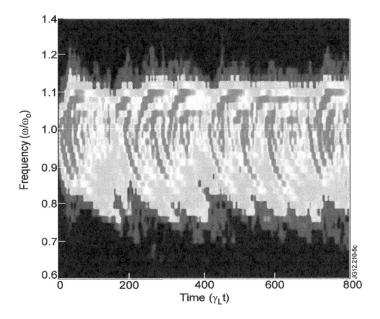

FIGURE 8.15   Non-linear HAGIS simulation of Alfvén instability in MAST #27177.

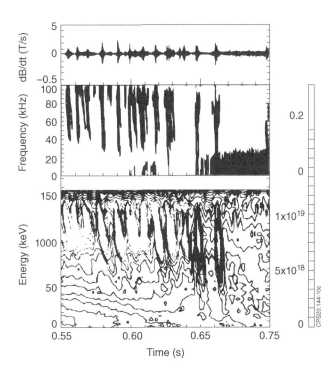

FIGURE 8.16   Top: Mirnov coil data; Middle: magnetic spectrogram; Bottom: Time evolution of tangential energetic neutral spectrum. The NPA viewing angle is set to 0°.

and with the beam pure drag-like relaxation model installed was performed to compare non-linear wave-particle dynamics with the experimentally observed in Figure 8.14. Figure 8.15 shows that the HAGIS modelling reproduces pretty well the characteristic spectrum observed in experiments, although the range of the frequency sweeping is not as large as this observed on MAST.

Finally, we note that the dominant transport mechanism for non-linear FS modes is convection of particles trapped in the wave field. Experimentally, the hole-clump formation and transport were observed for the first time with an NPA diagnostic on stellarator LHD [8.9]. Figure 8.16 shows how the flux of energetic beam ions sweeps in energy together with the chirping modes.

REFERENCES

1 H.L. Berk et al., *Phys. Rev. Lett.* 76 (1997) 1256.
2. A. Fasoli et al., *Phys. Rev. Lett.* 81 (1998) 5564.
3. R.F. Heeter et al., *Phys. Rev. Lett.* 85 (2000) 3177.
4. H.L. Berk et al., *Phys. Lett.* A234 (1997) 213.
5. M.K. Lilley et al., *Phys. Rev. Lett.* 102 (2009) 195003.
6. M.K. Lilley et al., *Phys. Plasma* 17 (2010) 092305.
7. N.V. Petviashvili, PhD Thesis, University of Texas at Austin (1999).
8. S.D. Pinches et al., *Comput. Phys. Comm.* 111 (1998) 133.
9. M. Osakabe et al., *Nucl. Fusion* 46 (2006) S911.

9 Alfvén Eigenmodes in "Advanced Tokamak" Plasmas

We now consider another type of AEs, different than TAE or other gap modes, with their frequencies locked to the equilibrium. These AEs were observed in discharges with reversed magnetic shear and were first called "Alfvén cascade" (AC) eigenmodes [9.1–9.3] due to special clustering of toroidal mode numbers of these excited modes. After theoretical identification of the modes and development of relevant models, the modes were also called reversed-shear AEs (RSAEs) [9.4]. We will be using both AC and RSAE abbreviations for these modes to reflect different papers published on this subject in the last 20 years.

A spontaneous improvement in fusion performance was observed in JET from the very first observations of ACs [9.5]. Figure 9.1 shows an example of such a case where the neutron rate in JET discharge was constant for about 1 s at fixed power of NBI and ICRH. However, this neutron yield almost doubles suddenly at $t \approx 7$ s without a change in either NBI or ICRH power. Simultaneously, a set of six AEs was detected with Mirnov coils starting at about $t \approx 7$ s. The modes start mostly at a frequency of ≈ 90 kHz, well below TAEs seen at ~180–240 kHz, and sweep in frequency up to ≈ 140–250 kHz, each branch having its own slope of df/dt and ending at its own frequency. The sweeping rate of the modes is rather slow indicating that it is likely to be associated with equilibrium change, for example, with q-profile evolution than with the non-linear wave-particle phenomena described in Chapter 8. Two questions arise: is the correlation between the improvement of fusion performance and the AEs observed just a coincidence? If not, can we use such correlation for scenario development? A sudden doubling of the neutron rate definitely deserves an investigation!

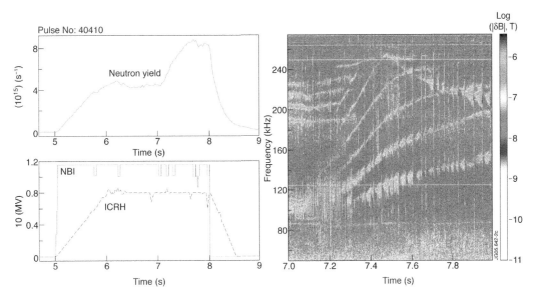

FIGURE 9.1 Left: Neutron yield, NBI and ICRH power in JET discharge #40410 with magnetic field B_T =3.4 T and plasma current I_p =3 MA. Right: ACs seen at the time of neutron rate increase.

9.1 ADVANCED TOKAMAK SCENARIOS AND INTERNAL TRANSPORT BARRIERS

Advanced tokamak (AT) scenarios aim at obtaining a high fusion performance plasma at significantly reduced inductive current while maximising off-axis bootstrap current and using current drive systems [9.6–9.8]. The AT scenarios often trigger internal transport barriers (ITBs), which are rather narrow radial regions in the plasma core, across which transport of plasma heat and/or mass are suppressed. The central region inside an ITB could deliver a very high fusion performance. In particular, the record in D-D neutron rate, that is, $S_n \approx 5.5 \times 10^{16} s^{-1}$ (pulse #40554), the JET record ion temperature, $T_i(0) \approx 40$ keV, and the ion temperature and pressure radial gradients, ≈ 150 keV/m and $\approx 10^6$ Pa/m (pulse #42940), were achieved in ITB discharges on JET with carbon wall [9.9]. To develop an ITB, the main heating power is applied during the inductive current ramp-up, as shown in Figure 9.2, before the current flat top is achieved. On JET, low hybrid current drive (LHCD) and/or low-power ICRH are often used additionally in the early ramp-up phase of discharges to adjust the q-profile if required.

When main NBI heating power is applied to the same AT scenario, but at different times in different discharges, for example, at $t_1, \ldots, t_4$, each of these times corresponds to a different inductive current, as shown in Figure 9.2.

In accordance with the diffusion of inductive current and current drive applied early, the safety factor evolves through a series of profiles $q(r,t)$ in the early pre-heating phase, so different "target" q-profiles correspond to different times $t_1, \ldots, t_4$ as Figure 9.3 illustrates.

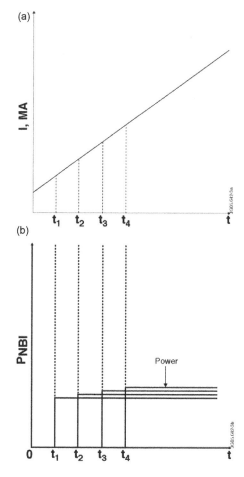

FIGURE 9.2 Schematic figures showing (a) the inductive current ramp-up in time, and (b) main NBI heating power applied at different times.

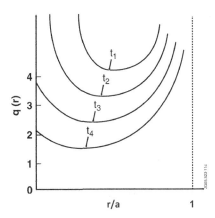

FIGURE 9.3 Schematic illustration of $q(r)$-profiles corresponding to different times $t_1,..., t_4$ of the current ramp-up phase shown in Figure 9.1.

When the main NBI heating is applied, plasma temperature increases strongly, so the resistive diffusion time increases and the evolution of $q(r, t)$ becomes slow. Essentially, this means that by selecting the time of main heating one picks up a "target" q-profile to be further explored for ITB development.

An ITB could spontaneously arise in AT plasmas with both positive and negative magnetic shear. However, triggering of an ITB in plasma with positive shear has a threshold in the input power, while triggering of an ITB in reversed-shear plasma has a very little or no threshold in the input power, and is much easier to develop. On JET, LHCD of low power (≤ 2 MW) was often employed before the main heating to obtain a reversed-shear (non-monotonic q-profile) magnetic configuration for further ITB development. Although LHCD applied was of low power, it was sufficient enough to affect significantly the early "embryo" state of the plasma in the very beginning of a discharge formation.

The ITB triggering has a very interesting and important correlation with integer values of the safety factor q [9.10,9.11]. In particular, ITBs are most often formed and sustained at the $q = 2$ or $q = 3$ magnetic flux surfaces in the case of positive shear. In the case of reversed shear, ITBs are easiest to form when $q_{min}(t)$ passes an integer value, which is most often $q_{min} = 2$.

9.2 ALFVÉN CASCADE (AC) EIGENMODES IN AT DISCHARGES WITH REVERSED MAGNETIC SHEAR

9.2.1 EXPERIMENTAL OBSERVATIONS

Experiments designed for generating ITBs in reversed-shear JET plasmas reveal discrete spectrum of Alfvén perturbations with predominantly upward frequency sweeping. These experiments are characterised by a hollow plasma current profile often facilitated by LHCD before the main heating power phase. The AEs observed are driven by ICRH-accelerated ions. Let us consider a typical example of observing the AEs in an ITB discharge [9.3]. Figure 9.4 shows a comparison of two very similar AT discharges on JET, in which the main heating power was applied at ~3.5 s during inductive current ramp-up (the current temporal evolution is shown by broken line). The only difference between these two discharges is LHCD of low power, ~2.5 MW, applied in one of the discharges, pulse #49382).

Figure 9.5a and b show magnetic fluctuation data measured by the magnetic pick-up coils in the Alfvén frequency range during the pre-heating phase of the two discharges in Figure 9.4. A comparison of Figure 9.4a and b reveals that the discrete spectra of AEs are very different in the two discharges. In the discharge without LHCD (pulse #49384), the ICRH-accelerated H-minority ions

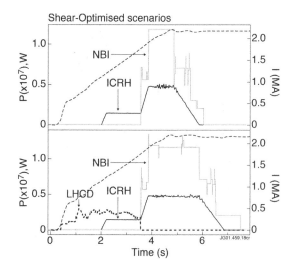

FIGURE 9.4 Power wave-forms of NBI (grey color), ICRH (black), and LHCD (black thick broken line) in two comparison JET discharges (pulses #49384 in the top and #49382 in the bottom) with $B_T = 2.6$ T and $I_P^{max} = 2.2$ MA. The heating power and LHCD are applied during current ramp-up phase when the current increases from $I_P = 1.1$ MA at $t = 2$ s to $I_P = 2.2$ MA at $t = 5$ s, see broken thin trace showing the current evolution. A non-monotonic q-profile is measured in the pulse #49382 with LHCD, while the pulse #49384 has monotonic q-profile.

excited usual TAEs, as shown in Figure 9.5a, whose frequency followed the increase of plasma current in time (shown by the black line in the left top corner of the figure). The comparison discharge with LHCD exhibits some AEs with the frequency sweeping below the TAE frequency, as shown in Figure 9.5b. These are AC eigenmodes, or RSAEs.

We note here that ACs were observed in discharge #49382 even when LHCD was switched off. This shows that ACs are associated with plasma equilibrium created by LHCD, rather than with LHCD itself. At the same time, because no ACs were observed without ICRH in JET discharges, we conclude that ICRH-accelerated ions are essential for ACs.

9.2.2 ANALYSIS OF THE EXPERIMENTAL OBSERVATIONS

The toroidal mode numbers of ACs shown in Figure 9.4b do not change for each branch of the perturbation as the relevant mode frequency sweeps upward. The frequency of the ACs starts from 40 to 90 kHz, that is, well below the TAE frequency observed in the comparison discharge (pulse #49384). During the cascade evolution, the frequency increases up to the frequency of the TAE-gap, so that the TAE-gap forms an "envelope" of ACs at their highest frequencies. The rate of increase in the AC frequency is proportional to the mode number n and modes of different n occur at different times, sometimes in isolation, and sometimes clustering with other modes of different n's. Interestingly, the temporal evolution of ACs shown in Figure 9.5(b) exhibits a characteristic pattern typical of plots of several branches of Alfvén continuum with different n's plotted on top of each other as, for example, in the CSCAS modelling part of Figure 7.9. This indicates that one needs to investigate in depth the Alfvén continuum as a function of time and different n's for explaining the ACs.

For each branch of the AC, the frequency changes on a time scale $\tau \cong 0.1 \div 0.5$ s, which is significantly longer than the time scale observed for non-linear "chirping" modes described in Chapter 7. On the other hand, the time scale of the frequency change is of the order of the current increase in time scale shown in Figure 9.4b. We conclude that the temporal evolution of plasma equilibrium, and especially the temporal evolution of plasma current and associated $q(r)$-profile, are important ingredients for explaining the frequency evolution of ACs. A direct measurement of the $q(r)$-profiles

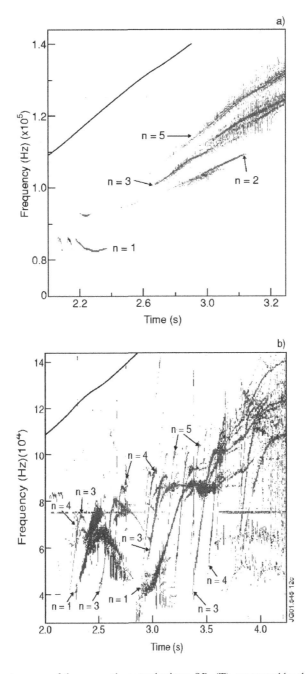

FIGURE 9.5 (a) Spectrogram of the magnetic perturbations, δB_P (T), measured by the Mirnov coils in JET discharge #49384. TAEs are observed at frequencies $f_{TAE} \cong 80$–200 kHz. Evolution of TAE frequency in time is caused by the plasma current increase shown with black trace in the left top corner. (b) Magnetic spectrogram in JET discharge #49382. Multiple branches of Alfvén cascades ranging from $n = 1$ to $n = 6$ are observed at frequencies well below TAE frequency range, $f_{AC} \cong 30$–100 kHz $\ll f_{TAE}$, with the frequency sweeping rate proportional to n. The black trace in the left top corner shows the plasma current increase [9.2, 9.3].

in the two comparison discharges was performed with motional stark effect (MSE) technique [9.12]. Figure 9.6 shows a dramatic difference in the $q(r)$-profiles: the discharge #49384 (no LHCD) had a monotonic $q(r)$-profile, while the comparison discharge #49382 (LHCD used) had a non-monotonic $q(r)$-profile with a strong negative magnetic shear in the plasma core.

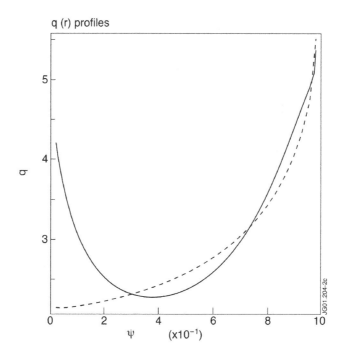

FIGURE 9.6 Safety factor profiles $q(\psi\_\mathrm{pol})$ reconstructed from EFIT with MSE for JET discharge #49384 at ~4.5 s (broken line) and #49382 at ~4.5 s (solid line).

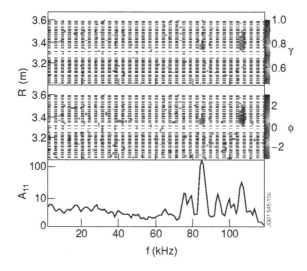

FIGURE 9.7 Cross-correlation spectrogram for the amplitude γ and the phase ϕ of an Alfvén cascade. The radial location of the perturbation at $R \approx 3.4$ m (which corresponds to $r/a \approx 0.4$) is inferred from the cross-correlation between the external magnetics and the 48-channel ECE diagnostic.

In some JET discharges with high ICRH power, internal plasma measurements with the electron cyclotron emission (ECE) and soft X-ray (SXR) diagnostics were also available for identifying the radial position of ACs. An example of such measurements is shown in Figure 9.7 for JET plasmas with non-monotonic $q(r)$ and ICRH power in excess of 7 MW (pulse #53494). This level of ICRH power in this discharge was much higher than the ≈ 2 MW power threshold needed to excite the ACs.

In the pulse with internal measurements, for a certain time interval of ~200 ms when two branches of ACs were observed at 85 kHz and 110 kHz, the cross-correlation analysis between the perturbed electron temperature, δT_e, and the magnetic perturbations, δB_{pol}, was performed, as shown in Figure 9.7. The perturbed electron temperature is measured with 48-channel ECE diagnostics with radially separated lines-of-sight, so that the radial mode structure in the plasma core could be determined. The two pictures on top of Figure 9.7 show the amplitude and phase of the cross-correlation integral $I(r)$ of the type

$$I(r) \propto \int \frac{\delta B_p \cdot \delta T_e(r)}{|\delta B_p| \cdot |\delta T_e(r)|} dt, \tag{9.1}$$

where the minor radius r is different for the different ECE channels and the time interval for the integration is 200 ms. From Figure 9.7 one can estimate the mode location on the low-field side of the torus as seen from the top. The two ACs, $n=3$ at $f \cong 85$ kHz and $n=6$ at $f \cong 110$ kHz, are found to be localised at major radius, ~3.4 m (magnetic axis is at 2.97 m), corresponding to $r/a \approx 0.4$. This is close to the zero magnetic shear region where the q-profile has its lowest value of q_{min}. It is interesting to note that the modes of significantly different frequencies and toroidal mode numbers are localised at nearly the same position in radius.

A similar technique of cross-correlation between SXR measurements (which also depends on electron temperature perturbation, δT_e) and magnetic perturbation, δB_{pol}, gives an estimate of the localisation of the AC in vertical direction because the X-ray camera has horizontal lines-of-sight. It is found that the mode at $f \cong 85$ kHz has localisation peaks at a vertical minor radius of ~60 cm, which is again close to the zero magnetic shear region (taking into account the elliptical cross-section of JET plasma). No clear correlation was found on SXR for the mode at $f \cong 110$ kHz. The measurements were repeated in this discharge 1 s later. For the branches of the Alfvén cascades seen at that time, $n=2$ at $f \cong 70$ kHz, $n=3$ at $f \cong 95$ kHz, and $n=4$ at $f \cong 120$ kHz, we found from the ECE measurements that the mode localisation is at ~3.3 m, that is, still at position close to the AC region in Figure 9.7. The X-ray measurements indicate that the vertical localisation of the modes is also close to the one measured at an earlier time.

Therefore, we conclude that AC localisation region is (1) close to the position of q_{min}; (2) it is the same for different mode numbers; and (3) the mode localisation does not change significantly on a time scale of ≈ 1 s.

9.2.3 Evolution of Alfvén Continuum due to Temporal Evolution of Non-Monotonic q-Profile

Because the ACs exist in reversed magnetic shear discharges and seem to be localised in the vicinity of the q_{min} magnetic flux surface, an investigation of Alfvén waves associated with the region surrounding the zero magnetic shear point, as well as study of the temporal evolution of the key equilibrium parameters in this region are necessary for interpreting ACs. We start by investigating the Alfvén continuum in the toroidal geometry and for plasma equilibrium with non-monotonic q-profile. Due to the non-monotonic q-profile, the shear Alfvén continuous frequency as a function of radius has extremum points, $r = r_0$, satisfying

$$\left. \frac{d\omega_A(r)}{dr} \right|_{r=r_0} = 0, \tag{9.2}$$

and so the Alfvén spectrum may contain discrete eigenvalues as in the cases of GAE and TAE.

The q-profile evolves during AC observations, so we need to investigate the Alfvén continuum as a function of radius and time. We calculate the continuum with the ideal MHD CSCAS code, which accounts for toroidal geometry. The temporal evolution of the Alfvén continuum frequency at the point of zero magnetic shear at q_{min} ("tip" of the Alfvén continuum) is of major interest. In a "cylindrical" limit, this evolution is described by:

$$\omega_A \left(r = r_{min}, t \right) = \left| \frac{m}{q_{min}(t)} - n \right| \cdot \frac{V_A(t)}{R_0}, \tag{9.3}$$

where r_{min} is the zero magnetic shear point, and $q_{min}(t)$ and $V_A(t)$ vary in time in accordance with the experiment. This behaviour is quite similar to what emerges from the toroidal CSCAS code except that the numerical code automatically switches the dominant m-number as the frequency approaches the TAE-gap frequency. Figure 9.8 shows the evolution from the CSCAS code of the value $\omega_A(t)$ for $n=1$ as $q_{min}(t)$ gradually decreases from 3 to 2.4. It follows from Figure 9.8 that during the entire evolution, the Alfvén continuum frequency as a function of radius has a local extremum point determined by (9.2), which is very close to the point of zero magnetic shear, q_{min}. During most of this time evolution, the Alfvén continuum has a local *maximum* at q_{min} except near the final point where q_{min} approaches $q_{min}=2.4$ (labelled 7). At low frequency, this local maximum of the frequency linearly increases as q_{min} decreases, but as the maximum frequency approaches the TAE-gap, the local maximum changes to the local minimum of the Alfvén continuum at q_{min} (marker 7 in Figure 9.8). During the subsequent decrease of q_{min} to 2, this minimum of the Alfvén continuum decreases to zero frequency as the rational magnetic surface is formed at $q_{min}=2$. On a longer time scale, the evolution of $n=1$ Alfvén continuum frequency as a function of q_{min} is shown in Figure 9.9 [9.2].

The repetition rate of the Alfvén continuum tip passing zero frequency is determined by the rate of plasma current increase and diffusion, which causes q_{min} to pass integer values during the interval when q_{min} decreases. During this scan, the highest frequency achieved by the local maximum of the

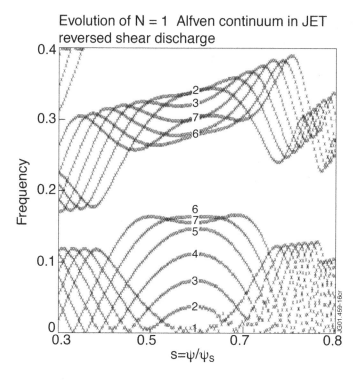

FIGURE 9.8   The CSCAS modelling: normalised frequency $\omega R_0 / V_A(0)$ of $n=1$ Alfvén continuum as a function of radius $s = \left(\psi_P / \psi_P^{edge} \right)^{1/2}$ for several values of q_{min} associated with evolution of $q_{min}(t)$ from $q_{min}=3$ down to $q_{min}=2.4$ in reversed-shear discharge. The sequence of Alfvén continuum "tips" corresponding to values $q_{min}=3$, 2.9, 2.8,..., 2.4 is shown by numbers 1,..., 7.

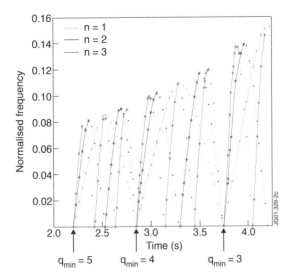

FIGURE 9.9   The CSCAS analysis showing temporal evolution of the normalised frequency $\omega_A(r_{min})R_0 / V_A$ at $q = q_{min}$ as $q_{min}(t)$ decreases in time. Mode numbers plotted are $n=1$, $n=2$, and $n=3$. Solid lines indicate times of local maximum of the Alfvén continuum, while broken lines indicate times of local minimum of the Alfvén continuum [9.2, 9.3].

Alfvén continuum is bounded by the frequency of the TAE-gap. Because the frequency of TAE-gap is inversely proportional to q_{min}, the envelope of the highest frequencies is roughly proportional to the plasma current increase during the time of observation. If now we consider the temporal evolution of the Alfvén continuum frequency at q_{min} for higher toroidal mode numbers, $n=2$ and $n=3$, we observe a similar characteristic pattern of the intermittent maximum and minimum of Alfvén continuum at decreasing q_{min}. However, the rate of the frequency sweeping and the repetition rate for the higher mode numbers are higher than for $n=1$, for example, the $n=2$ mode frequency sweeps approximately twice as fast, and this mode passes zero frequency not only at integer values of q_{min} but at half-integer values as well.

A direct comparison of the experimental data shown in Figure 9.4b and the CSCAS modelling in Figure 9.8 suggests that the following formula describes the frequency sweeping of the Alfvén cascades:

$$\omega(t) = \left(\frac{m}{q_{min}(t)} - n \right) \frac{V_A}{R_0} + \Delta\omega, \tag{9.4}$$

where $\Delta\omega$ is an off-set frequency determined by possible Doppler shift in toroidally rotating plasma, by toroidal coupling corrections of the Alfvén continuum, etc.

Taking into account the correspondence between the temporal evolution of the Alfvén continuum tip and the experimentally observed data, our basic hypothesis concerning ACs is that these modes are similar to a global Alfvén eigenmode, whose frequency is close to the frequency where the localised shear Alfvén wave has an extremum as a function of radius. However, in the standard theory of the GAE, in cylindrical geometry, the extremum of the local Alfvén wave is a minimum. However, we see on comparing Figure 9.4b and 9.8 that only upward sweeping ACs are observed experimentally, which correspond to the maximum in Alfvén continuum, in contrast to GAE.

9.2.4   THEORETICAL EXPLANATION OF ACS

Taking into account all the essential results of our previous analysis and the experimental data, a discrete spectrum of modes associated with non-monotonic $q(r)$-profiles was found in Refs.

[9.2,9.13]. Within the MHD description of shear Alfvén perturbations and the drift kinetic description of energetic particles, the following equation was derived for a single cylindrical harmonic of the mode electrostatic potential:

$$
\frac{d}{dr}\left[\frac{\omega^2}{\bar{V}_A^2} - \frac{1}{R_0^2}\left(n - \frac{m}{q}\right)^2\right] r \frac{d}{dr}\Phi_m
$$

$$
- \frac{m^2}{r^2} r \left[\frac{\omega^2}{\bar{V}_A^2} - \frac{1}{R_0^2}\left(n - \frac{m}{q}\right)^2\right]\Phi_m
$$

$$
= \frac{m^2}{r^2} r \frac{\omega^2}{\bar{V}_A^2}\left(2\frac{\varepsilon^2 + 2\varepsilon\Delta'}{(2qn - 2m)^2 - 1}\right)\Phi_m + \frac{4\pi e}{cB} m\Phi_m
$$

$$
\times \frac{d}{dr}\left[\omega\langle n_h\rangle - \frac{1}{eR_0}\left(n - \frac{m}{q}\right)\langle j_{\|h}\rangle\right].
\tag{9.5}
$$

Here, the left-hand side consists of the usual shear Alfvén terms with Alfvén velocity calculated for a local mass density but for the on-axis magnetic field. The right-hand side represents two different effects causing the formation of an AC eigenmode: the first term $\sim \varepsilon^2$ results from the toroidal geometry and is important for plasmas with weak reversed shear, while the terms in the second bracket with flux-averaged density and current of energetic (hot) ions result from electrons compensating energetic ion charge when drift orbits of energetic ions are much larger than the mode width. The very large orbit width compared to the radial mode width is clearly visible on comparing Figures 2.7 and 9.7. By looking for an eigenmode with eigenfrequency very close to the Alfvén continuum "tip" frequency (9.3) in the vicinity of $q_{\min}$ and taking into account the zero shear condition at the position $r = r_0$ of $q_{\min}$, we expand the parallel wave-vector around $q_{\min}(r_0) \equiv q_0$ as

$$
\frac{1}{R_0^2}\left(n - \frac{m}{q(r)}\right)^2 \cong \frac{1}{R_0^2}\left(n - \frac{m}{q_q}\right)^2
$$

$$
+ \frac{mq''(r)(r - r_0)^2}{q_0^2 R_0^2}\left(n - \frac{m}{q_0}\right)
\tag{9.6}
$$

while the frequency is supposed to be close to $\omega_0 = k_{\|}(r_0) \cdot V_A(r_0)$:

$$
\omega^2 - \omega_0^2 \approx 2\omega_0(\omega - \omega_0).
\tag{9.7}
$$

Then, the starting Eq. (9.5) reduces to the form

$$
\frac{d}{dx}D_m\frac{d\Phi_m}{dx} - D_m\Phi_m = 2\frac{\omega_0^2 R_0^2 q_0^2}{\bar{V}_A^2}\frac{\varepsilon_0^2 + 2\varepsilon_0\Delta_0'}{(2q_0 n - 2m)^2 - 1}\Phi_m
$$

$$
+ \frac{\omega_0^2 R_0^2 q_0^2}{\bar{V}_A^2}\frac{\omega_{ch}}{\omega_0}\frac{r_0}{m}\left(\frac{d\langle\rho_h\rangle}{\rho dr}\right)_{r=r_0}\Phi_m
\tag{9.8}
$$

where

$$D_m(x) = \left[2(\omega - \omega_0)\omega_0 \frac{q_0^2 R_0^2}{\bar{V}_A^2} - (q_0 n - m) \frac{r_0^2 q_0''}{q_0 m} x^2 \right] \tag{9.9}$$

This second-order differential equation can be represented in the form of Schrodinger equation [9.2,9.13], with the existence of a discrete eigenvalue (corresponding to the AC mode existence) possible if

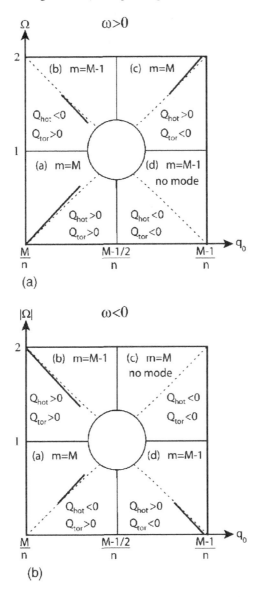

FIGURE 9.10 Schematic plot of normalised AC frequencies $\Omega \equiv \omega/\omega_{TAE}$ for two selected poloidal mode numbers M and $M-1$. The different quadrants correspond to different signs of Q_{hot} and Q_{tor}. Solid lines indicate possible AC modes, dashed lines indicate the Alfvén continuum, the circles indicate the TAE frequency regions where the single harmonic mode approximation fails. The relevant values of m are shown in each quadrant[*].

$$Q_{eff} \equiv Q_{hot} + Q_{tor} > 1/4 \tag{9.10}$$

where

$$Q_{hot} = \left(\frac{d\langle n_h \rangle}{dr} \right)_{r=r_0} \frac{4\pi e R_0^2 q_0^3}{c B r_0 q_0''} \frac{\omega_0}{(m - n q_0)}$$

$$\equiv \omega_0^2 \frac{q_0^2 R_0^2}{V_A^2 (m - n q_0)} \frac{q_0}{r_0^2 q_0''} \left[\left(-\frac{r}{\rho} \frac{d\langle \rho_h \rangle}{dr} \right)_{r=r_0} \frac{\omega_{ch}}{\omega_0} \right] \tag{9.11}$$

$$Q_{tor} = m \omega_0^2 \frac{2 q_0^2 R_0^2}{V_A^2 (m - n q_0)} \frac{q_0}{r_0^2 q_0''} \frac{\varepsilon_0 (\varepsilon_0 + 2\Delta_0')}{\left[1 - (2 q_0 n - 2m)^2 \right]} \tag{9.12}$$

Figure 9.10 shows schematically what types of ACs may exist below and above TAE-gap depending on the contributions from (9.11) and (9.12). These values could be positive or negative depending on the value of q_0 and the sign of AC frequency.

REFERENCES

1. S.E. Sharapov et al., *Phys. Lett. A* **289** (2001) 127.
2. H.L. Berk et al., *Phys. Rev. Lett.* **87** (2001) 185002.
3. S.E. Sharapov et al., *Phys. Plasmas* **9** (2002) 2027.
4. R. Nazikian et al., *Fusion Energy*, Proc. 20th Int. Conf. Vilamoura (IAEA, Vienna, 2004), paper EX/5-1.
5. S.E. Sharapov et al., *Nucl. Fusion* **46** (2006) S868.
6. T.S. Taylor, *Plasma Phys. Control. Fusion* **39** (1997) B47.
7. C.D. Challis et al., *Plasma Phys. Control. Fusion* **43** (2001) 861.
8. R.C. Wolf, *Plasma Phys. Control. Fusion* **45** (2003) R1.
9. C. Gormezano et al., *Phys. Rev. Lett.* **80** (1998) 5544.
10. E. Joffrin et al., *Nucl. Fusion* **43** (2003) 1167.
11. M.E. Austin et al., *Phys. Plasmas* **13** (2006) 082502.
12. N.C. Hawkes et al., *Phys. Rev. Lett.* **87** (2001) 115001.
13. B.N. Breizman et al., *Phys. Plasmas* **10** (2003) 3649.

10 Alfvén Spectroscopy

Alfvén Eigenmodes driven by energetic particles are routinely seen in tokamak plasmas heated by ICRH and/or by NBI. Measurements of such AEs with low amplitude represent an attractive form of MHD spectroscopy, which solves the inverse problem of deducing plasma equilibrium parameters from the observed spectrum of the MHD perturbations [10.1–10.3]. AEs are more attractive than other MHD modes as they are robust and numerous, studied in-depth, and do not cause significant degradation of plasma or the fast ion confinement as long as AE amplitudes remain small. We call this branch of MHD spectroscopy "Alfvén spectroscopy" to underline the essential use of discrete spectra of Alfvén waves.

In this chapter, we demonstrate several important applications of Alfvén spectroscopy to tokamak plasmas. First, we will discuss typical observations of core-localised TAEs in sawtoothing discharges with ICRH [10.4–10.7]. In the presence of ICRH-accelerated ions stabilising sawteeth, the sawtooth period increases significantly, and the time derivative dT_e / dt decreases, or, in some cases, becomes negative just before the next sawtooth crash. Such long-period sawteeth are called "monster sawteeth" as they cause a very significant drop in central electron temperature. Evolution of the safety factor $q(r)$ preceding to such sawtooth crashes and the value of the safety factor after the crashes are some of the most important issues in tokamak physics [1.3]. The use of ICRH generating tails of very energetic ions in such plasmas is a natural source of excitation of AEs. Therefore, it is possible to apply Alfvén spectroscopy to deduce the information on the temporal evolution of the $q(r)$-profile throughout the sawtooth cycle, as described in Section 10.1.

Second, the use of Alfvén spectroscopy has become routine in the development of advanced tokamak scenarios because the detection of ACs and TAEs helps to identify the existence of reversed magnetic shear and provide accurate data on the evolution of $q_{min}(t)$ [10.2]. It was established for JET reversed-shear discharges that the timing of $q_{min}(t)$ passing an integer surface coincides with a spontaneous improvement of electron heat confinement, called an ITB triggering event. Development of an ITB in reversed-shear AT scenarios with the use of ACs is discussed in Section 10.2.

Third, in Section 10.3, we discuss the use of Alfvén spectroscopy in determining D:T ratio in the plasma core [10.3]. The fusion performance strongly depends on the D:T ratio, and diagnostics of this ratio are of the highest priority for the next-step burning plasmas. The D:T ratio at the plasma edge is measured from the intensity of D_α and T_α spectral lines; however, the D:T ratio in the core of plasma with fuelling of tritium or deuterium NBI could be different and requires a specific diagnostic tool for the plasma core.

10.1 ALFVÉN EIGENMODES, SAWTEETH, AND THE $q(r)$-PROFILE

The sawtooth is a robust phenomenon in tokamaks [1.3]. The sawtooth cycle consists of an abrupt crash in the central electron temperature followed by a slow reheat phase. In the sawtooth crash, the magnetic field "frozen-in" law breaks down leading to a reconnection of the magnetic field and a rapid radial transport causing a flattening of the temperature, density, and current profiles. The plasma reheat phase occurs after the crash, and is characterised by an increase in the central electron temperature and an inward diffusion of plasma current.

In the studies of sawteeth in tokamak plasmas, the safety factor $q(r)$ plays a crucial role. However, it is difficult to measure the $q(r)$-profile accurately in experiment, especially when the magnetic shear is low. In this section, we review how the information on the $q(r)$-profile could be obtained from the AEs observed before and after the sawtooth crashes.

One of the earliest examples of applying Alfvén spectroscopy to a monster sawtooth was obtained in JT-60U (Japan) experiments with the second harmonic ICRH of hydrogen minority [10.4]. Figure 10.1 shows a magnetic spectrogram with many AEs detected, and the main heating and plasma parameters in a JT-60U discharge with $B_T = 3.35$ T, $I_P = 2.6$ MA, $R_{0} = 3.47$ m, $a = 1.03$ m. One can see from the magnetic spectrogram that TAEs preceding the sawtooth crash at $t \approx 9.95$ s appear in a sequence of decreasing toroidal mode numbers from $n = 12$ at $t \approx 9.3$ s to $n = 4$ at $t \approx 9.75$. We reasonably assume (this is confirmed later with proper modelling) that the plasma preceding the sawtooth crash has the magnetic surface $q = 1$, and the TAEs are localised inside the $q = 1$ magnetic surface; therefore, this succession of TAEs excited before the crash is associated with a slow decrease in the q-profile in the plasma core. Hence, we can establish from (D.4) the following time sequence of the q-values corresponding to the appearance of TAE with every n: $q = 11.5/12 = 0.958$ ($n = 12$), followed by 0.955 ($n = 11$), 0.95 ($n = 10$), 0.944 ($n = 9$), 0.9375 ($n = 8$), 0.929 ($n = 7$), 0.917 ($n = 6$), 0.9 ($n = 5$), 0.875 ($n = 4$). Each q-value here corresponds to TAE localisation region off-axis (where the profile of ICRH-accelerated energetic ions peaks), but the on-axis, q-values should not differ much due to the weak magnetic shear inside the $q = 1$ radius.

Immediately after the crash, TAEs appear with toroidal mode numbers from $n = 12$ to $n = 14$. However, the sequence of the toroidal numbers is reversed compared with that of the TAEs before the crash. The increasing toroidal mode numbers indicate that the TAEs are now located in the

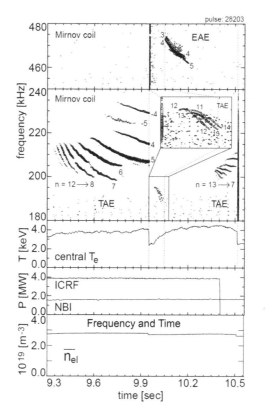

FIGURE 10.1 Typical discharge showing various groups of AEs before and after sawtooth crashes. After the crash, weak TAE activity is observed first (inset) followed by the excitation of EAEs.

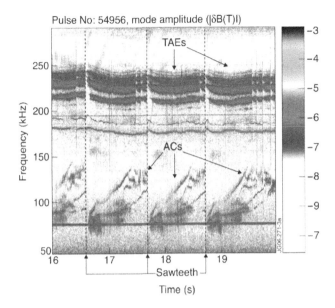

FIGURE 10.2   ACs and TAEs in JET #54956 with monster sawteeth.

region with $q > 1$, and the sequence of TAE-related q-values is 1.04 ($n = 12$), 1.038 ($n = 13$), and 1.036 ($n = 14$).

After the crash, but somewhat later, a bunch of ellipticity-induced AEs (EAEs) appear at more than double the TAE frequency. All the modes appear at once although they have different toroidal mode numbers, $n = 3,..., 7$, and the frequency separation between the modes is very small at less than 3 kHz. The modelling [10.4] shows that the EAEs are located at the $q = 1$ surface, and hence, the appearance and evolution of these modes could provide the information on the time of the $q = 1$ appearance and its temporal evolution.

Similar interplay between AEs and sawteeth was confirmed on other tokamaks. In addition, it was found that some sawtooth reconnection events cause a weakly negative magnetic shear in the plasma centre [10.5]. This conclusion was made from observing grand-ACs (Alfvén grand cascades with all n's excited at once). The grand-ACs are usually used for diagnosing integer $q_{min} > 1$ needed for developing ITB scenarios. However, some JET discharges with high T_e values and profiles of electron temperature strongly peaked within a very broad $q = 1$ radius show the presence of grand-ACs repeatedly occurring after each sawtooth crash, as shown in Figure 10.2. In these discharges, ACs have been used for diagnosing q_{min} in monster sawteeth with weakly reversed magnetic shear. From the observations of the ACs, it was found that the crashes occur at $q_{min} \sim 0.85$, while after the crashes, $q_{min} > 1$ was deduced.

Similar observations on the Alcator C-Mod tokamak [10.6] allowed two important parameters, q_{min} and its radial position, to be deduced from the comparison of the AC spectra and MHD modelling of ACs. These studies provided valuable constraints for the modelling of the sawtooth cycle on this machine.

Investigation of the sawtooth cycle with the high sensitivity O-mode interferometry on JET has shown the relation between ACs observed just after some crashes, as well as TAEs preceding the crashes [10.7]. Figure 10.3 shows JET plasmas with nearly steady-state ICRH and many sawteeth, some of which were monster sawteeth. Figure 10.4 shows how the pattern of AEs evolves over time from ACs to TAEs, with the slow q-profile evolution. In the case shown, some of the TAEs originate from ACs and represent the same mode, but at different positions of the swept frequency.

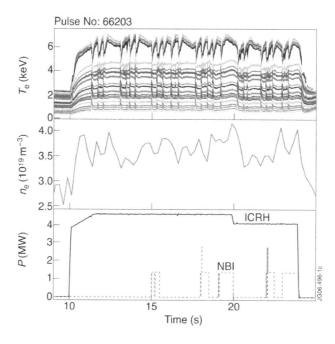

FIGURE 10.3   Top: Electron temperature T_e measured with multi-channel ECE diagnostics at different radii on JET. Middle: Electron density $n_e(0)$ measured with LIDAR diagnostics. Bottom: ICRH power wave-form shown in solid line and diagnostic NBI shown in dashed line.

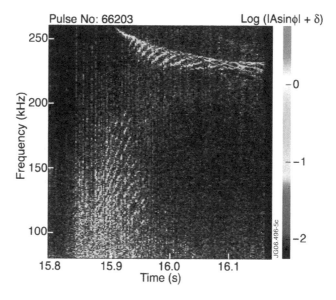

FIGURE 10.4   Spectrogram of high-frequency density perturbations measured with the O-mode interferometry at frequency 50.47 GHz just before a monster sawtooth crash at $t = 16.15$ s. Alfvén cascades are seen first in the frequency range of 100–200 kHz. Tornado modes (TAEs with decreasing n's inside the $q = 1$ radius) are seen in the TAE frequency range around 200–240 kHz.

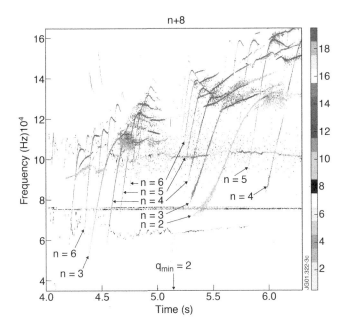

FIGURE 10.5   Magnetic spectrogram showing ICRH-driven Alfvén Cascade eigenmodes in shear-reversed JET discharge #53488 (B_T=2.5 T, I_P=2.2 MA) around the time of q_{min}=2 appearing.

10.2   TRIGGERING OF ITBS BY LOW-ORDER RATIONAL q_{min} AND ALFVÉN GRAND CASCADES

In this section, we first consider in detail the spectroscopic pattern of an AC in reversed-shear tokamak discharges. Figure 10.5 shows the pattern consisting of many ACs with different toroidal mode numbers n.

In accordance with the theory/modelling presented in Chapter 9, the time appearance of every AC is associated with q_{min} passing through the relevant rational number. The Alfvén spectroscopy in this case is based on the clustering of ACs with different n's in time [10.2], that is,

$n = 1$ ACs occur when $q_{min} = 1, 2, 3...$;
$n = 2$ ACs occur when $q_{min} = 1, 3/2, 2...$;
$n = 3$ ACs occur when $q_{min} = 1, 4/3, 5/3...$

The bunch of ACs, in which all n's are present at once is called "grand cascade." It occurs when q_{min} passes an integer value. Figure 10.5 shows how the ACs appear one-by-one just before the time when q_{min} passes an integer value of 2, while they all emerge at once just after such event at $t \sim 5.15$ s. Because of the easily recognisable characteristic pattern of the grand cascades, it is easy to diagnose experimentally the times when q_{min} passes integer values in JET shear-reversed discharges with ICRH or NBI exciting ACs. By comparing the grand cascade times and the times of ITB formation in shear-reversed JET plasmas, it was established that these two times of very different origin correlate [10.8].

Further investigation has shown that low rational values of q_{min} in plasmas with non-monotonic $q(r)$-profiles cause improved electron confinement seen as an increase by ~10%–20% in the central electron temperature [10.9]. We call this spontaneous improvement of the confinement an

"ITB triggering event." If no significant heating is applied to the plasma at that time, the ITB triggering event gradually disappears in ~100–200 ms. However, if a significant main heating is applied just before the ITB triggering event, a full-scale ITB could be developed. Therefore, we conclude that the ITB triggering event is a necessary but not sufficient condition for producing an ITB. The timing of the main heating power determines whether the ITB triggering event can end up in a full-scale ITB. Figure 10.6 illustrates how the ITB triggering event looks like in the same JET discharge, as shown in Figure 10.5.

The correlation between Alfvén grand cascades and the ITB triggering events was investigated in JET experiments with different types of pre-heating, that is, with LHCD, ICRH+LHCD, ICRH only, NBI only, and with pellets. The correlation was found to exist in JET plasmas with densities up to ~5×10^{19}m^{-3}, indicating that the timing of ITB triggering from the Grand-ACs may facilitate

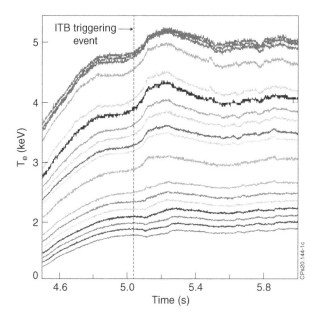

FIGURE 10.6 Electron temperature traces measured with multi-channel ECE diagnostics. A sudden increase in the slope is observed at $t \approx 5.1$ s, indicating a sudden improvement in electron confinement.

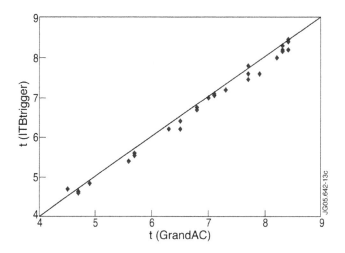

FIGURE 10.7 Correlation between the times of ITB triggering times and grand cascades.

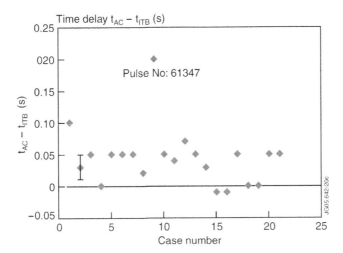

FIGURE 10.8 Time delay between Grand-ACs and ITB triggering events for a set of discharges with interferometry diagnostic used for the AC detection.

ITB scenario development in machines with high densities. Several plasma conditions were investigated too,

- $1.5 < I_p < 2.2$ MA
- $2.45 < B_T < 3.4$ T
- $3 < P_{total} < 17$ MW.

The results of these studies are presented in Figure 10.7. Similar observations have been made on DIII-D [10.10].

For using this kind of Alfvén spectroscopy in practice, a discharge with ICRH power is run at the beginning of a specific experimental session. All Grand-ACs are identified in such discharge, thus providing the time sequence of ITB triggering events and $q_{min}(t) = integer$ surfaces appearing in this particular plasma. The timing of the main heating power is decided then for the rest of the session for making an ITB associated with one of the integer values of q_{min}.

With the improved time resolution of AE detection (see Figure D.4 in the Appendix D), it also became possible to investigate more accurately whether the ITB triggering event is preceding Grand-AC marking $q_{min}(t) = integer$ or it occurs just after that. Figure 10.8 strongly suggests that the ITB triggering events occur before the appearance of the integer surface in the plasma. The most plausible theory that may explain such sequence of events is the theory associated with the depletion in time of rational magnetic surfaces of drift waves turbulence [10.11,10.12] just before q_{min} reaches an integer value, rather than the presence of an integer q_{min} value itself.

10.3 PLASMA MASS DETERMINATION FROM ACTIVE AE MEASUREMENTS

The square root dependence of Alfvén velocity on the plasma mass has suggested from the very beginning of MHD spectroscopy the diagnostic potential of the AE frequency measurements for determining the plasma isotopic mix [10.1]. After the discovery of the gap AEs, including TAEs and EAEs, this diagnostic potential was demonstrated during JET DT campaign with the use of active TAE diagnostic described in Section 7.1 [10.3]. The measurements of TAE frequency were performed in a series of nearly identical JET discharges with varied D:T concentrations [10.13].

Figure 10.9 presents the results obtained with three different diagnostics including the active TAE for two JET discharges with different range of the D:T ratio. The figure suggests a good

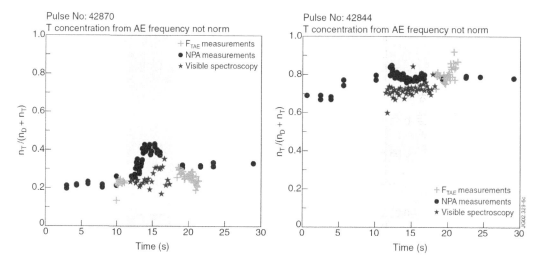

FIGURE 10.9 Estimate of D-T ratio from the measured frequency of an externally excited $n = 1$ TAE compared with visible spectroscopic data (intensity ratio of T_α to D_α lines) and with the neutral particle analyser results[*].

agreement with edge spectroscopic measurements, although with a somewhat different time evolution. This method may be useful in a reactor where the same plasma configuration will repeatedly be employed. Direct measurements are possible in real time for similar equilibria, with a time resolution of ~30 ms, which is the time needed to scan the frequency of the probe wave across a TAE resonance. The overall accuracy depends on the reproducibility of the equilibrium. When the equilibrium is significantly varied from shot to shot along with the isotopic mix, one must rely upon a full theoretical analysis.

REFERENCES

1. J.P. Goeldbloed et al., *Plasma Phys. Control. Fusion* **35** (1993) B277.
2. S.E. Sharapov et al., *Phys. Lett. A* **289** (2001) 127.
3. A. Fasoli et al., *Plasma Phys. Control. Fusion* **44** (2002) B159.
4. G.J. Kramer et al., *Nucl. Fusion* **40** (2000) 1383.
5. S.E. Sharapov et al., *Fusion Energy* Proc. 21st Int. Conf. (IAEA, Chengdu, 2006) paper EX/P6-19.
6. E.M. Edlund et al., *Phys. Rev. Lett.* **102** (2009) 165003.
7. P. Sandquist et al., *Phys. Plasmas* **14** (2007) 122506.
8. E. Joffrin et al., *Nucl. Fusion* **43** (2003) 1167.
9. S.E. Sharapov et al., *Nucl. Fusion* **46** (2006) S868.
10. M.E. Austin et al., *Phys. Plasmas* **13** (2006) 082502.
11. A.D. Beklemishev and W. Horton, *Phys. Fluids* **B4** (1992) 200.
12. X. Garbet et al., *Nucl. Fusion* **43** (2003) 975.
13. P.R. Thomas et al., *Phys. Rev. Lett.* **80** (1998) 5548.

[*] Reproduced from [A. Fasoli et al., *Plasma Phys. Control. Fusion* 44 (2002) B159], with the permission of IOP Publishing.

Appendix A for Chapter 3

A.1 COLLISIONAL RELAXATION OF ANISOTROPIC ENERGETIC BEAM IONS

Consider a temporal evolution of energetic beam ion distribution function, $f(v_x, v_y, v_z)$, in the velocity space due to Coulomb collisions with plasma electrons and ions. The plasma considered is homogeneous, and an axial symmetry of the beam distribution function is assumed during the evolution:

$$f(v_x, v_y, v_z) = f(v, \vartheta); v = \sqrt{v_\parallel^2 + v_\perp^2}; \vartheta = \tan^{-1}(v_\perp / v_\parallel);$$

$$v_\perp = \sqrt{v_x^2 + v_y^2}; v_\parallel = v_z. \tag{A.1}$$

The beam evolution can be described by the Fokker–Planck equation for hot ion distribution function $f(v, \vartheta)$ with initial velocity V_0 for the typical range of hot ion velocities between ion and electron thermal velocities, $v_i \ll V_0 \ll v_e$. Neglecting self-collisions between hot ions, the linear Fokker–Planck equation can be written as [A.1,A.2]:

$$\frac{\partial f}{\partial t} = \frac{V_0^3}{\tau v^2} \left\{ \frac{\partial}{\partial v} \left[\frac{V_0^2 a(v)}{2v} \frac{\partial f}{\partial v} + b(v) f \right] + \frac{c(v)}{V_0} \cdot \frac{1}{\sin \vartheta} \frac{\partial}{\partial \vartheta} \left[\sin \vartheta \frac{\partial f}{\partial \vartheta} \right] \right\} - vf + pF, \tag{A.2}$$

where

$$a(v) = \frac{T_e}{E_0} \left\{ \tilde{Z}_2 + \frac{4}{3\sqrt{\pi}} \frac{m_0}{m_e} \left(\frac{v}{v_e} \right)^3 \right\}; b(v) = \tilde{Z}_1 + \frac{4}{3\sqrt{\pi}} \frac{m_0}{m_e} \left(\frac{v}{v_e} \right)^3;$$

$$c(v) = Z_{\text{eff}} \frac{V_0}{2v} + \frac{2}{3\sqrt{\pi}} \frac{V_0}{v_e} = Z_{\text{eff}} \frac{V_0}{2v} \left(1 + \frac{4}{3\sqrt{\pi}} \frac{v}{v_e} \right); \tau = \frac{1}{\pi\sqrt{2}} \frac{E_0^{3/2} m_0^{1/2}}{Z_b^2 e^4 n_e L};$$

$$\tilde{Z}_1 = \frac{m_0}{n_e} \sum_i \frac{Z_i^2 n_i}{m_i}; \tilde{Z}_2 = \frac{m_0}{n_e T_e} \sum_i \frac{Z_i^2 n_i T_i}{m_i}; Z_{\text{eff}} = (1/n_e) \sum_i Z_i^2 n_i;$$

L is the Coulomb logarithm, m_0 is the beam ion mass, E_0, V_0 are the initial energy and speed of fast ions, respectively, and the last two terms in the right-hand side of (A.2) represent sink and source of the energetic ions, respectively. The source term satisfies the normalisation

$$2\pi \int_0^\infty v^2 dv \int_0^\pi F(v - V_0; \vartheta) d\vartheta = 1. \tag{A.3}$$

A comparison of the terms proportional to a, b, and c shows that for the highest energy range of the beam ions, $v \sim V_0$, the dominant relaxation effect is the slowing down of the beam ions due to Coulomb collisions with thermal electrons and ions. This effect is determined by the term proportional to $b(v)$. The effect of beam scattering in the pitch angle ϑ is the next in the ordering. This effect proportional to $c(v)$ is mostly determined by beam collisions with thermal ions, while the electron contribution is small as V_0/v_e. Finally, the velocity diffusion term proportional to $a(v)$ is the weakest, with a small factor T_e/E_0 in front of it.

To analyse (A.2), we drop the small term proportional to $a(v)$ and represent both the source and the distribution function of the energetic ions via a series of Legendre polynomials [A.2]:

$$F(v, \vartheta) = \sum_l F_l(v - V_0) \cdot P_l(\cos\vartheta), \tag{A.4}$$

$$f(v, \vartheta) = \sum_l f_l(v) \cdot P_l(\cos\vartheta), \tag{A.5}$$

$$v = v(v). \tag{A.6}$$

Here, $P_l(\cos\vartheta)$ are the Legendre polynomials satisfying

$$\frac{(z^2 - 1) dP_l(z)}{dz} = lzP_l(z) - lP_{l-1}(z), \tag{A.7}$$

so that (A.2) can be represented as set of one-dimensional first-order differential equations for the functions $\kappa_l \equiv b(v) f_l$:

$$\frac{\partial \kappa_l}{\partial t} - \frac{V_0^3 b(v)}{\tau v^2} \frac{\partial \kappa_l}{\partial v} = -\kappa_l \left\{ \frac{V_0^3 l(l+1) c(v)}{\tau v^2 V_0} + v(v) \right\} + p b(v) F_l(v - V_0), \tag{A.8}$$

resulting in characteristic relations

$$dt = -\frac{\tau v^2 dv}{V_0^3 b(v)} = \frac{d\kappa_l}{-\kappa_l \left\{ \dfrac{V_0^3 l(l+1) c(v)}{\tau v^2 V_0} + v(v) \right\} + p b(v) F_l(v - V_0)}. \tag{A.9}$$

The evolution of the beam ion speed in time can be found from (A.9):

$$\int_{V_0}^{v} \frac{\tau v^2 dv}{V_0^3 b(v)} = -\int_0^t dt, \tag{A.10}$$

resulting in time evolution

$$v = V_I \left(\left(\tilde{Z}_1 + \left(\frac{V_0}{V_I}\right)^3 \right) \exp\left(-\frac{V_I}{V_0} \frac{t}{3\tau} \right) - \tilde{Z}_1 \right)^{1/3}, \tag{A.11}$$

where

$$V_I \equiv v_e \left(\frac{3\sqrt{\pi}}{4} \cdot \frac{m_e}{m_0} \right)^{1/3}.$$

Expression (A.11) gives $v = V_0$ at $t = 0$ and the following simple expression

$$v \approx V_0 \left(1 - \frac{V_I}{V_0} \frac{t}{3\tau} \right). \tag{A.12}$$

The beam slows down to $v \sim V_I$ at

$$t \sim 3\tau \frac{V_0}{V_I} \ln\left(\left(\tilde{Z}_1 + \left(\frac{V_0}{V_I}\right)^3\right) \middle/ \left(\tilde{Z}_1 + 1\right)\right).$$ (A.13)

The l-th component of the velocity distribution function can be obtained from (A.9). After straight-forward lengthy calculations, the following general form of κ_l can be found for arbitrary source and sink:

$$\kappa_l(v, t) = -\frac{p\tau}{V_0^3} \int_{V_0}^{v} \xi^2 d\xi F_l(\xi - V_0) \exp\left(-\frac{\tau}{V_0^3} \int_{v}^{\xi} \eta^2 d\eta \frac{G_l(\eta)}{b(\eta)}\right) +$$

$$+ \frac{p\tau}{V_0^3} \int_{V_0}^{W(t, v)} \xi^2 d\xi F_l(\xi - V_0) \exp\left(-\frac{\tau}{V_0^3} \int_{v}^{\xi} \eta^2 d\eta \frac{G_l(\eta)}{b(\eta)}\right),$$ (A.14)

where

$$G_l(\eta) \equiv \frac{V_0^3 l(l+1) c(v)}{\tau v^2 V_0} + \nu(v),$$ (A.15)

and

$$W(v, t) = V_0 \left(\left(\alpha^3 + \left(\frac{v}{V_0}\right)^3\right) \cdot \exp\left(\frac{t}{\tau_0}\right) - \alpha^3\right)^{1/3}$$ (A.16)

satisfies $W(v, t) \geq v$,

$$W(v, t = 0) = v,$$

$$W(v, t \to \infty) \to \infty.$$ (A.17)

Here,

$$\alpha \equiv \left(\frac{3\sqrt{\pi}}{4}\right)^{2/3} \left(\frac{m_0}{m_e}\right)^{1/3} \frac{T_e}{E_0},$$

$$\tau_0 \equiv \tau \frac{\sqrt{\pi}}{4} \frac{m_e}{m_0} \left(\frac{v_e}{V_0}\right)^3.$$

REFERENCES TO APPENDIX A

[A.1]   Yu.N. Dnestrovkii and D.P. Kostomarov, *Mathematical modelling of plasma*, Nauka, Moscow (1993) (in Russian).
[A.2]   H.L. Berk et al., *Nucl. Fusion* **15** (1975) 819.

Appendix B for Chapter 4
Curvilinear Coordinates in Toroidal Geometry

B.1 COORDINATE TRANSFORMS

We begin with Cartesian coordinates (x, y, z) with the basis of unit vectors $(\boldsymbol{e}_x, \boldsymbol{e}_y, \boldsymbol{e}_z)$

In this case, representation of a vector has the form:

$$A = x\boldsymbol{e}_x + y\boldsymbol{e}_y + z\boldsymbol{e}_z$$

with components of a vector:

$$x = (\boldsymbol{e}_y \times \boldsymbol{e}_z) \cdot A$$

$$y = (\boldsymbol{e}_z \times \boldsymbol{e}_x) \cdot A$$

$$z = (\boldsymbol{e}_x \times \boldsymbol{e}_y) \cdot A$$

If one transforms to another coordinate system (α, β, γ), then

$$A = \alpha\boldsymbol{e}_\alpha + \beta\boldsymbol{e}_\beta + \gamma\boldsymbol{e}_\gamma,$$

with

$$\alpha = x\frac{\partial x}{\partial \alpha} + y\frac{\partial y}{\partial \alpha} + z\frac{\partial z}{\partial \alpha},$$

$$\beta = x\frac{\partial x}{\partial \beta} + y\frac{\partial y}{\partial \beta} + z\frac{\partial z}{\partial \beta},$$

$$\gamma = x\frac{\partial x}{\partial \gamma} + y\frac{\partial y}{\partial \gamma} + z\frac{\partial z}{\partial \gamma}.$$

Let us generalise the maths above to curvilinear coordinates (ξ^1, ξ^2, ξ^3) with the covariant basis $(\nabla\xi^1, \nabla\xi^2, \nabla\xi^3)$ and the covariant representation of a vector: $A = A_1\nabla\xi^1 + A_2\nabla\xi^2 + A_3\nabla\xi^3$. The covariant components of the vector take the form:

$$A_1 = J(\nabla\xi^2 \times \nabla\xi^3) \cdot A;$$

$$A_2 = J(\nabla\xi^3 \times \nabla\xi^1) \cdot A;$$

$$A_3 = J(\nabla\xi^1 \times \nabla\xi^2) \cdot A$$

where Jacobian is given by $J = \left[\left(\nabla \xi^1 \times \nabla \xi^2 \right) \cdot \nabla \xi^3 \right]^{-1}$

By substituting the Jacobian, one obtains:

$$A = \frac{\left(\nabla \xi^2 \times \nabla \xi^3 \right) \cdot A}{\left(\nabla \xi^1 \times \nabla \xi^2 \right) \cdot \nabla \xi^3} \nabla \xi^1 + \frac{\left(\nabla \xi^3 \times \nabla \xi^1 \right) \cdot A}{\left(\nabla \xi^1 \times \nabla \xi^2 \right) \cdot \nabla \xi^3} \nabla \xi^2 + \frac{\left(\nabla \xi^1 \times \nabla \xi^2 \right) \cdot A}{\left(\nabla \xi^1 \times \nabla \xi^2 \right) \cdot \nabla \xi^3} \nabla \xi^3$$

If one transforms to another coordinate system $\left(\tilde{\xi}^1, \tilde{\xi}^2, \tilde{\xi}^3 \right)$, then

$$A = \tilde{A}_1 \nabla \tilde{\xi}^1 + \tilde{A}_2 \nabla \tilde{\xi}^2 + \tilde{A}_3 \nabla \tilde{\xi}^3,$$

with

$$\tilde{A}_i = \sum_j A_j \frac{\partial \xi^j}{\partial \tilde{\xi}^i}$$

We can also introduce a contravariant basis $\left(J \left(\nabla \xi^2 \times \nabla \xi^3 \right), J \left(\nabla \xi^3 \times \nabla \xi^1 \right), J \left(\nabla \xi^1 \times \nabla \xi^2 \right) \right)$

with contravariant components: $A^1 = A \cdot \nabla \xi^1, A^2 = A \cdot \nabla \xi^2, A^3 = A \cdot \nabla \xi^3$

The contravariant representation of a vector is then obtained in the form:

$$A = A \cdot \nabla \xi^1 \frac{\left(\nabla \xi^2 \times \nabla \xi^3 \right)}{\left(\nabla \xi^1 \times \nabla \xi^2 \right) \cdot \nabla \xi^3} + A \cdot \nabla \xi^2 \frac{\left(\nabla \xi^3 \times \nabla \xi^1 \right)}{\left(\nabla \xi^1 \times \nabla \xi^2 \right) \cdot \nabla \xi^3} + A \cdot \nabla \xi^3 \frac{\left(\nabla \xi^1 \times \nabla \xi^2 \right)}{\left(\nabla \xi^1 \times \nabla \xi^2 \right) \cdot \nabla \xi^3}$$

Metric tensor is a relationship between covariant and contravariant components:

$$A^i = \sum_j A_j \nabla \xi^j \cdot \nabla \xi^i = \sum_j g^{ij} A_j,$$

where the contravariant components of tensor g are given by

$$g^{ij} = \nabla \xi^i \cdot \nabla \xi^j = \frac{\partial \xi^i}{\partial x} \cdot \frac{\partial \xi^j}{\partial x} + \frac{\partial \xi^i}{\partial y} \cdot \frac{\partial \xi^j}{\partial y} + \frac{\partial \xi^i}{\partial z} \cdot \frac{\partial \xi^j}{\partial z}$$

and covariant components of tensor g are given by the inverse matrix:

$$g_{ij} = \left[g^{ij} \right]^{-1} = \frac{\partial x}{\partial \xi^i} \cdot \frac{\partial x}{\partial \xi^j} + \frac{\partial y}{\partial \xi^i} \cdot \frac{\partial y}{\partial \xi^j} + \frac{\partial z}{\partial \xi^i} \cdot \frac{\partial z}{\partial \xi^j}$$

$$\text{Det} \left(g_{ij} \right) \equiv g = J^2$$

Volume element in the general geometry case takes the form:

$$d^3 x = dx dy dz = J d\xi^1 d\xi^2 d\xi^3$$

and the length element takes the form:

$$(ds)^2 = (dx)^2 + (dy)^2 + (dz)^2 = \sum_{i,j=1} g_{ij} d\xi^i d\xi^j$$

In vector calculus, divergence, which is a vector operator that operates on a vector field producing a scalar field giving the quantity of the vector field's source at each point, becomes:

$$\nabla \cdot A = \frac{1}{J} \frac{\partial}{\partial \xi^i} \left(J A^i \right)$$

while the curl, which is a vector operator describing the infinitesimal rotation of a vector field in three-dimensional Euclidean space, takes the form:

$$(\nabla \times A)^i = \nabla \xi^i \cdot \sum_{j,m} \nabla \xi^j \times \left(\frac{\partial}{\partial \xi^j} \nabla \xi^m A_m \right) = \frac{1}{J} \sum_{j,k} \varepsilon_{ijk} \frac{\partial A_k}{\partial \xi^j}$$

where ε_{ijk} is an anti-symmetric unit matrix whose only non-zero components are

$$\varepsilon_{123} = \varepsilon_{231} = \varepsilon_{312} = 1; \varepsilon_{132} = \varepsilon_{213} = \varepsilon_{321} = -1$$

B.2   SHAFRANOV COORDINATES FOR TOROIDAL PLASMA

We start from the coordinate system employed by Shafranov for analysing plasma equilibrium in toroidal geometry (Figure B.1).

The following expressions relate the different sets of coordinates:

$$R = R_0 + r \cos \vartheta - \Delta(r) = \sqrt{x^2 + y^2};$$

$$\phi = -\zeta = \tan^{-1}(y/x);$$

$$Z = r \sin \vartheta;$$

with the following inversed relations:

$$x = R \cos \phi = R \cos \zeta;$$

$$y = R \sin \phi = R \sin \zeta$$

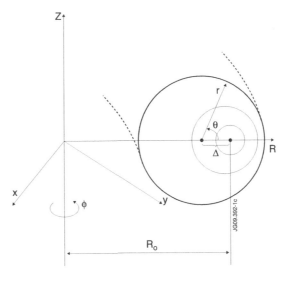

FIGURE B.1   Cartesian coordinates (x, y, Z), polar coordinates (R, φ, Z), and toroidal coordinates $(r, \theta, \varsigma = -\varphi)$ used for describing toroidal plasma equilibrium. Here, R_0 is the major radius of the magnetic axis, and Δ is the Shafranov shift (the distance between R_0 and the geometric major radius).

The 3×3 matrix for transformations from (x, y, Z) to $\xi^1 = r, \xi^2 = \vartheta, \xi^3 = \zeta$ has the following elements:

$$\frac{\partial x}{\partial r} = \left(\frac{\partial R}{\partial r}\right)\cos\zeta \qquad \frac{\partial x}{\partial \vartheta} = \left(\frac{\partial R}{\partial \vartheta}\right)\cos\zeta \qquad \frac{\partial x}{\partial \zeta} = -R\sin\zeta$$

$$\frac{\partial y}{\partial r} = -\left(\frac{\partial R}{\partial r}\right)\sin\zeta \qquad \frac{\partial y}{\partial \vartheta} = -\left(\frac{\partial R}{\partial \vartheta}\right)\sin\zeta \qquad \frac{\partial y}{\partial \zeta} = -R\cos\zeta$$

$$\frac{\partial Z}{\partial r} = \sin\vartheta \qquad\qquad \frac{\partial Z}{\partial \vartheta} = r\cos\vartheta \qquad\qquad \frac{\partial Z}{\partial \zeta} = 0$$

so that the covariant components of the metric tensor

$$g_{ij} = \frac{\partial x}{\partial \xi^i} \cdot \frac{\partial x}{\partial \xi^j} + \frac{\partial y}{\partial \xi^i} \cdot \frac{\partial y}{\partial \xi^j} + \frac{\partial z}{\partial \xi^i} \cdot \frac{\partial z}{\partial \xi^j}$$

are calculated easily, for example, for g_{11}:

$$g_{11} = \left(\frac{\partial R}{\partial r}\right)^2 + \sin^2\vartheta = \left(\cos\vartheta - \Delta'\right)^2 + \sin^2\vartheta = 1 - 2\Delta'\cos\vartheta + \Delta'^2$$

to give the following matrix:

$$g_{ij} = \begin{pmatrix} 1 - 2\Delta'\cos\vartheta + \Delta'^2 & r\Delta'\sin\vartheta & 0 \\ r\Delta'\sin\vartheta & r^2 & 0 \\ 0 & 0 & R^2 \end{pmatrix}$$

with the following Jacobian relevant to Shafranov coordinates:

$$J_S = \sqrt{\det\left(g_{ij}\right)} = rR \cdot \left(1 - \Delta'\cos\vartheta\right) \cong rR_0 \cdot \left(1 + \left(\frac{r}{R_0} - \Delta'\right)\cdot\cos\vartheta\right)$$

Here, we used the large aspect ratio ordering

$$\Delta/R_0 \cong \varepsilon^2 \ll \varepsilon = \left(r/R_0\right) \ll 1$$

The contravariant metric elements can be found from the general expression:

$$g^{ij} = \left(g_{ij}\right)^{-1} = \frac{1}{J^2}\begin{pmatrix} g_{22} \cdot g_{33} & -g_{12} \cdot g_{33} & 0 \\ -g_{12} \cdot g_{33} & g_{11} \cdot g_{33} & 0 \\ 0 & 0 & g_{11} \cdot g_{22} - g_{12}^2 \end{pmatrix}$$

These contravariant elements are:

$$g^{11} = \nabla r \cdot \nabla r = 1 + 2\Delta'\cos\vartheta$$

$$g^{12} = \nabla r \cdot \nabla\vartheta = -\frac{1}{r}\Delta'\sin\vartheta$$

$$g^{22} = \nabla\vartheta \cdot \nabla\vartheta = \frac{1}{r^2}$$

$$g^{33} = \nabla\zeta \cdot \nabla\zeta = \frac{1}{R^2}$$

B.3   FLUX-TYPE COORDINATES IN TORUS

For further investigations of plasma stability and waves, we need to introduce a safety factor

$$q = \frac{\boldsymbol{B} \cdot \nabla\zeta}{\boldsymbol{B} \cdot \nabla\vartheta}$$

which plays a major role. A significant simplification of the maths associated with plasma stability and waves can be achieved if the safety factor is *independent* of the poloidal angle variable ϑ. As the density of magnetic field lines is higher at the inner side of the torus (where equilibrium magnetic field is higher), a different poloidal coordinate than the usual poloidal angle used in Shafranov coordinates is required. Due to the explicit connection to the flux of equilibrium magnetic field, the coordinates with such improved poloidal angle variable can be called flux-type coordinates. One can search for a desired set of flux-type coordinates $(r_f; \vartheta_f; \zeta_f)$ related to Shafranov coordinates $(r_S; \vartheta_S; \zeta_S)$ via

$$r_f = r_S; \quad \vartheta_f = \vartheta_f(r_f, \vartheta_S); \quad \zeta_f = \zeta_S$$

where the expression for ϑ_f can be found from the condition $\partial q/\partial\vartheta_f = 0$. For this, we split $\boldsymbol{B}$ into poloidal and toroidal parts,

$$\boldsymbol{B} = \nabla\zeta \times \nabla\psi + I(\psi)\nabla\zeta$$

and obtain

$$(\nabla\zeta)^2 = (1/R)^2, \quad \boldsymbol{B} \cdot \nabla\zeta_f = I(\psi)/R^2$$

$$\nabla\psi = (d\psi/dr)\nabla r, \quad \boldsymbol{B} \cdot \nabla\vartheta_f = \frac{d\psi}{dr} \cdot \left(\frac{1}{\nabla r_f \times \nabla\vartheta_f \cdot \nabla\zeta_f}\right)^{-1} = \frac{\psi'}{J_f}$$

where $J_f = 1/(\nabla r_f \times \nabla\vartheta_f \cdot \nabla\zeta_f)$ is the *flux-type Jacobian*. We now explicitly require

$$q = \frac{\boldsymbol{B} \cdot \nabla\zeta}{\boldsymbol{B} \cdot \nabla\vartheta} = q(\psi),$$

and obtain

$$J_f = \frac{R^2(r,\vartheta) \cdot q(\psi)\psi'}{I(\psi)},$$

with the ϑ-dependence involved only in R^2. The ratio of the Jacobians derived above gives a differential equation for the relevant poloidal angle variables:

$$\frac{\partial\vartheta_f}{\partial\vartheta_S} = \frac{J_S}{J_f} \propto \frac{1 + ((r/R_0) - \Delta')\cos\vartheta_S}{1 + 2(r/R_0)\cos\vartheta_S} \cong 1 - \left(\frac{r}{R_0} + \Delta'\right)\cos\vartheta_S.$$

One can then find the flux-type poloidal angle variable via Shafranov angle:

$$\vartheta_f = \vartheta_S - \left(\frac{r}{R_0} + \Delta'\right)\sin\vartheta_S + C$$

Here, $C=0$ as ϑ_f must change by 2π whenever ϑ_S does. Thus,

$$J_f = J_S\left(\frac{\partial\vartheta_f}{\partial\vartheta_S}\right)^{-1} = \frac{rR_0\left(1+\left((r/R_0)-\Delta'\right)\cos\vartheta\right)}{1-\left((r/R_0)+\Delta'\right)\cos\vartheta} \cong \frac{rR^2}{R_0}.$$

Inverting the expression for $\vartheta_f\left(r_f,\vartheta_S\right)$ gives

$$\vartheta_S = \vartheta_f + \left(\frac{r}{R_0} + \Delta'\right)\sin\vartheta_f$$

and we can write using Taylor expansions:

$$\cos\vartheta_S \cong \cos\vartheta_f + \left((r/R_0)+\Delta'\right)\sin\vartheta_f \cdot \left(-\sin\vartheta_f\right) = \cos\vartheta_f - \eta(r)\left(1-\cos 2\vartheta_f\right);$$

$$\sin\vartheta_S \cong \sin\vartheta_f + \left((r/R_0)+\Delta'\right)\sin\vartheta_f \cdot \left(+\cos\vartheta_f\right) = \sin\vartheta_f + \eta(r)\sin 2\vartheta_f;$$

where

$$\eta(r) \equiv \frac{1}{2}\left(\frac{r}{R_0} + \Delta'\right)$$

We can now rewrite Shafranov coordinates to obtain the flux-type coordinates:

$$R = R_0 + r\cos\vartheta_f - \Delta(r) - r\eta(r)\left(1-\cos 2\vartheta_f\right);$$

$$\phi = -\zeta;$$

$$Z = r\sin\vartheta_f + r\eta(r)\sin 2\vartheta_f.$$

From this set of coordinates, we obtain the relevant flux-type Jacobian

$$J_f = \sqrt{\det\left(g_{ij}^f\right)} \cong rR\sqrt{1+2\frac{r}{R_0}\cos\vartheta_f} \cong \frac{rR^2}{R_0},$$

and for the flux-type coordinates we obtain the covariant matrix (put $\vartheta_f = \vartheta$ here):

$$g_{ij}^f = \begin{pmatrix} 1-2\Delta'\cos\vartheta & r\left((r/R_0)\sin\vartheta + (r\Delta')'\sin\vartheta\right) & 0 \\ r\left((r/R_0)\sin\vartheta + (r\Delta')'\sin\vartheta\right) & r^2\left(1+4\eta\cos\vartheta + 4\eta^2\right) & 0 \\ 0 & 0 & R^2 \end{pmatrix}$$

The contravariant metric elements are:

$$g^{11} = \nabla r \cdot \nabla r = 1 + 2\Delta' \cos \vartheta$$

$$g^{12} = \nabla r \cdot \nabla \vartheta = -\frac{1}{r}\left(\frac{r}{R_0} + (r\Delta')'\right)\sin \vartheta$$

$$g^{22} = \nabla \vartheta \cdot \nabla \vartheta = \frac{1}{r^2}\left(1 - 2\left(\frac{r}{R_0} + \Delta'\right)\cos \vartheta\right)$$

$$g^{33} = \nabla \zeta \cdot \nabla \zeta = \frac{1}{R^2} = \frac{1}{R_0^2}\left(1 - 2\frac{r}{R_0}\cos \vartheta\right)$$

Appendix C for Chapter 6
Analytical Theory of TAE

C.1 DERIVATION OF TAE EQUATIONS

We start from the main governing equation (4.12) of ideal MHD

$$\boldsymbol{B} \cdot \nabla \left(\frac{\boldsymbol{J} \cdot \boldsymbol{B}}{B^2} \right) + \nabla \cdot \left(4\pi\rho \frac{\boldsymbol{B} \times \dfrac{d\boldsymbol{V}}{dt}}{B^2} \right) + \nabla \cdot \left(\frac{\boldsymbol{B} \times \nabla P}{B^2} \right) = 0 \tag{C.1}$$

and apply the linearization procedure

$$\boldsymbol{J} = \boldsymbol{J}_0 + \delta\boldsymbol{J}, \quad \boldsymbol{B} = \boldsymbol{B}_0 + \delta\boldsymbol{B}, \quad \boldsymbol{V} = \delta\boldsymbol{v}, \quad P = P_0 + \delta P, \tag{C.2}$$

which gives the following expressions:

$$\boldsymbol{B} \cdot \nabla \left(\frac{\boldsymbol{J} \cdot \boldsymbol{B}}{B^2} \right) \rightarrow \delta\boldsymbol{B} \cdot \nabla \left(\frac{\boldsymbol{J}_0 \cdot \boldsymbol{B}_0}{B_0^2} \right) +$$

$$\boldsymbol{B}_0 \cdot \nabla \left(\frac{1}{B_0^2} \left(\boldsymbol{B}_0 \cdot \delta\boldsymbol{J} + \boldsymbol{J}_0 \cdot \delta\boldsymbol{B} - 2\frac{\boldsymbol{J}_0 \cdot \boldsymbol{B}_0}{B_0^2} (\delta\boldsymbol{B} \cdot \boldsymbol{B}_0) \right) \right)$$

$$\nabla \cdot \left(\rho \frac{\boldsymbol{B} \times \dfrac{d\boldsymbol{V}}{dt}}{B^2} \right) \rightarrow \nabla \cdot \left(\frac{\rho}{B_0^2} \left(\boldsymbol{B}_0 \times \frac{d\delta\boldsymbol{v}}{dt} \right) \right)$$

$$\nabla \cdot \left(\frac{\boldsymbol{B} \times (\nabla P)}{B^2} \right) \rightarrow$$

$$\rightarrow \nabla \cdot \left(\frac{\boldsymbol{B}_0 \times \nabla\delta P + \delta\boldsymbol{B} \times \nabla P_0}{B_0^2} - 2\frac{\boldsymbol{B}_0 \times \nabla P_0}{B_0^4} (\delta\boldsymbol{B} \cdot \boldsymbol{B}_0) \right) \tag{C.3}$$

The last term in (C.3) associated with equilibrium pressure is zero as

$$\nabla \cdot \left(\boldsymbol{B}_0 \times \nabla P_0 \right) = B_0^2 \nabla \cdot \boldsymbol{J}_0 - (\boldsymbol{B}_0 \cdot \boldsymbol{J}_0) \nabla \cdot \boldsymbol{B}_0 = 0$$

By substituting all terms from (C.3) into (C.1), we obtain

$$\delta\boldsymbol{B} \cdot \nabla\left(\frac{\boldsymbol{J}_0 \cdot \boldsymbol{B}_0}{B_0^2}\right) + \boldsymbol{B}_0 \cdot \nabla\left(\frac{\boldsymbol{B}_0 \cdot \delta\boldsymbol{J}}{B_0^2}\right) + \boldsymbol{B}_0 \cdot \nabla\left(\frac{\boldsymbol{J}_0 \cdot \delta\boldsymbol{B}}{B_0^2}\right) -$$

$$-2\boldsymbol{B}_0 \cdot \nabla\left(\frac{(\boldsymbol{J}_0 \cdot \boldsymbol{B}_0)(\delta\boldsymbol{B} \cdot \boldsymbol{B}_0)}{B_0^4}\right) + \nabla \cdot \left(\frac{\rho}{B_0^2}\left(\boldsymbol{B}_0 \times \frac{d\delta\boldsymbol{v}}{dt}\right)\right) +$$

$$+\nabla \cdot \left(\frac{\boldsymbol{B}_0 \times (\nabla\delta P) + \delta\boldsymbol{B} \times \nabla P_0}{B_0^2}\right) = 0, \tag{C.4}$$

Two of the terms here give

$$\boldsymbol{B}_0 \cdot \nabla\left(\frac{\boldsymbol{J}_0 \cdot \delta\boldsymbol{B}}{B_0^2}\right) + \nabla \cdot \left(\frac{\delta\boldsymbol{B} \times \nabla P_0}{B_0^2}\right) = \ldots = \boldsymbol{J}_0 \cdot \nabla\left(\frac{\delta\boldsymbol{B} \cdot \boldsymbol{B}_0}{B_0^2}\right) \tag{C.5}$$

with the use of the equilibrium equation.

The linearized governing equation for a perturbed plasma takes the following form:

$$(\nabla \cdot \boldsymbol{J})^{(1)} = \boldsymbol{B}_0 \cdot \nabla\left(\frac{\delta\boldsymbol{J} \cdot \boldsymbol{B}_0}{B_0^2}\right) + \delta\boldsymbol{B} \cdot \nabla\left(\frac{\boldsymbol{J}_0 \cdot \boldsymbol{B}_0}{B_0^2}\right) - 2\boldsymbol{B}_0 \cdot \nabla\left(\frac{(\boldsymbol{J}_0 \cdot \boldsymbol{B}_0)(\delta\boldsymbol{B} \cdot \boldsymbol{B}_0)}{B_0^4}\right)$$

$$+\nabla \cdot \left(\frac{\rho}{B_0^2}\left(\boldsymbol{B}_0 \times \frac{d\delta\boldsymbol{v}}{dt}\right)\right) + \nabla \cdot \left(\frac{\boldsymbol{B}_0 \times \nabla\delta P}{B_0^2}\right) + \boldsymbol{J}_0 \cdot \nabla\left(\frac{\delta\boldsymbol{B} \cdot \boldsymbol{B}_0}{B_0^2}\right) = 0 \tag{C.6}$$

We introduce scalar and vector potentials for the electromagnetic waves of Alfvénic type, so the perturbed fields of the wave are

$$\delta\boldsymbol{E} = -\nabla\phi - \frac{\partial}{\partial t}\delta\boldsymbol{A} \tag{C.7}$$

$$\delta\boldsymbol{B} = \nabla \times \delta\boldsymbol{A}$$

Shear Alfvén waves satisfy the condition of zero parallel perturbed fields, that is,

$$\delta B_{\parallel} \equiv \frac{\delta\boldsymbol{B} \cdot \boldsymbol{B}_0}{B_0} = 0$$

so that only parallel component of the vector potential is involved:

$$\delta\boldsymbol{A} = \delta A_{\parallel}\frac{\boldsymbol{B}_0}{B_0} \tag{C.8}$$

Furthermore, as these waves have no parallel electric field, the scalar and vector potentials are related via

$$\delta E_{\parallel} \equiv \frac{\delta\boldsymbol{E} \cdot \boldsymbol{B}_0}{B_0} = -\frac{\boldsymbol{B}_0 \cdot \nabla\phi}{B_0} - \frac{\partial}{\partial t}\frac{\boldsymbol{B}_0 \cdot \delta\boldsymbol{A}}{B_0} = 0 \tag{C.9}$$

The cross-field perturbed velocity of plasma is simply

$$\delta\boldsymbol{v} = \boldsymbol{v}_E = \frac{1}{B_0^2}(\delta\boldsymbol{E}_{\perp} \times \boldsymbol{B}_0) \tag{C.10}$$

Using the Amperes law, we find

$$\delta J = -\frac{1}{\mu_0}\left(\nabla^2\left(\delta A_\parallel \frac{\boldsymbol{B}_0}{B_0}\right) - \nabla\left(\nabla\cdot\left(\delta A_\parallel \frac{\boldsymbol{B}_0}{B_0}\right)\right)\right) \tag{C.11}$$

For simplicity, by considering a small-β limit (so that $\delta P \rightarrow 0$.), we obtain

$$\boldsymbol{B}_0 \cdot \nabla\left(\frac{\delta\boldsymbol{J}\cdot\boldsymbol{B}_0}{B_0^2}\right) + \delta\boldsymbol{B}\cdot\nabla\left(\frac{\boldsymbol{J}_0\cdot\boldsymbol{B}_0}{B_0^2}\right) + \nabla\cdot\left(\frac{\rho}{B_0^2}\left(\boldsymbol{B}_0\times\frac{d\delta\boldsymbol{v}}{dt}\right)\right) = 0 \tag{C.12}$$

As the waves considered are extended along equilibrium magnetic field, $k_\parallel \ll k_\perp$, we obtain

$$\boldsymbol{B}\cdot\nabla\left(\frac{1}{B}\nabla_\perp^2\left(\boldsymbol{b}\cdot\nabla\phi\right)\right) + \nabla\cdot\left(\frac{\omega^2}{v_A^2}\nabla_\perp\phi\right) = 0 \tag{C.13}$$

This equation describes shear Alfvén waves in low-β plasmas. We reduce this three-dimensional equation to one-dimensional coupled equations by assuming solution in the form

$$\varphi(r,\vartheta,\zeta,t) = \exp(-i\omega t + in\zeta)\sum_m \varphi_m(r)\exp(-im\vartheta) + c.c. \tag{C.14}$$

Where $c.c.$ stands for "complex conjugate." For the equilibrium magnetic field in the form

$$\boldsymbol{B} = \nabla\zeta\times\nabla\psi + I(\psi)\nabla\zeta = \left(\nabla\psi = \frac{d\psi}{dr}\nabla r\right) = \psi'\nabla\zeta\times\nabla r + I(\psi)\nabla\zeta \tag{C.15}$$

we obtain

$$\nabla_\perp^2\phi = \frac{1}{J}\left(\frac{\partial}{\partial r}\left(Jg^{rr}\frac{\partial\phi}{\partial r} + Jg^{r\theta}\frac{\partial\phi}{\partial\theta}\right) + \frac{\partial}{\partial\theta}\left(Jg^{\theta\theta}\frac{\partial\phi}{\partial\theta} + Jg^{r\theta}\frac{\partial\phi}{\partial r}\right)\right) \tag{C.16}$$

$$\boldsymbol{b}\cdot\nabla\phi = \frac{\boldsymbol{B}\cdot\nabla\phi}{B} = \frac{1}{B}\left(\frac{B_\varphi}{R}\frac{\partial\phi}{\partial\zeta} + \frac{B_\theta}{r}\frac{\partial\phi}{\partial\theta}\right) =$$

$$= \frac{B_\theta}{rB}\left(\frac{\partial}{\partial\theta} + q\frac{\partial}{\partial\zeta}\right)\phi \equiv \left(\phi \sim e^{in\zeta}\right) \cong \frac{1}{qR}\left(\frac{\partial}{\partial\theta} + inq\right)\phi \tag{C.17}$$

where the metric components should be taken from the flux-type expressions given in Appendix B. Note that the parallel wave vector is given by

$$ik_{\parallel m,n}\phi \equiv \boldsymbol{b}\cdot\nabla_\parallel\phi_{mn} \cong \left(\phi \sim e^{i(n\zeta-m\theta)}\right) \cong$$

$$\cong \frac{1}{qR}(-im + inq)\phi \Rightarrow k_{\parallel m,n} \cong \frac{1}{qR}(nq - m) \tag{C.18}$$

The bending energy term takes the following form:

$$\boldsymbol{B}\cdot\nabla\left(\frac{1}{B}\nabla_\perp^2\left(\boldsymbol{b}\cdot\nabla\phi\right)\right) = \frac{I}{qR^2}\left(\frac{\partial}{\partial\zeta} + inq\right).$$

$$\frac{R}{IJ}\left(\frac{\partial}{\partial r}\left(Jg^{rr}\frac{\partial}{\partial r} + Jg^{r\theta}\frac{\partial}{\partial\theta}\right) + \frac{\partial}{\partial\theta}\left(Jg^{\theta\theta}\frac{\partial}{\partial\theta} + Jg^{r\theta}\frac{\partial}{\partial r}\right)\right). \tag{C.19}$$

$$\frac{1}{qR}\left(\frac{\partial}{\partial\theta} + inq\right)\phi$$

Considering the first term as an example, we obtain the following expression for this term:

$$
\text{First term, R.H.S.} = \frac{I}{qR_0^2}\left(1 - \frac{2r}{R_0}\cos\theta\right)\left(\frac{\partial}{\partial\theta} + inq\right) \cdot
$$

$$
\cdot\frac{1}{rI}\left(1 - \frac{r}{R_0}\cos\theta\right)\frac{\partial}{\partial r} rR_0\left(1 + 2\left(\frac{r}{R_0} + \Delta'\right)\cos\theta\right) \cdot \tag{C.20}
$$

$$
\cdot\frac{\partial}{\partial r}\frac{1}{qR_0}\left(1 - \frac{r}{R_0}\cos\theta\right)\left(\frac{\partial}{\partial\theta} + inq\right)\phi.
$$

We substitute $\varphi(r,\vartheta) = \sum_m \varphi_m(r)\exp(-im\vartheta)$, and use $\cos\vartheta = \left(e^{i\vartheta} + e^{-i\vartheta}\right)/2$ to pick the m-th element of the Fourier decomposition

$$
\langle Q\rangle_m = \frac{1}{2\pi}\int_0^{2\pi} Q e^{-im\theta}\,\mathrm{d}\theta \tag{C.21}
$$

to obtain the following expression for the first term:

$$
\langle\text{R.H.S. First Term}\rangle_m =
$$

$$
= \frac{1}{qrR_0^2}(-im + inq)\frac{\partial}{\partial r} rR_0\frac{\partial}{\partial r}\frac{1}{qR_0}(-im + inq)\phi_m
$$

$$
-\sum_\pm \frac{I}{qR_0^2}\frac{r}{R_0}(-i(m\pm 1) + inq)\frac{1}{rI}\frac{\partial}{\partial r} rR_0\frac{\partial}{\partial r}\frac{1}{qR_0}(-i(m\pm 1) + inq)\phi_{m\pm 1} +
$$

$$
-\sum_\pm \frac{I}{2qR_0^2}(-im + inq)\frac{1}{rI}\frac{r}{R_0}\frac{\partial}{\partial r} rR_0\frac{\partial}{\partial r}\frac{1}{qR_0}(-i(m\pm 1) + inq)\phi_{m\pm 1} +
$$

$$
+\sum_\pm \frac{I}{qR_0^2}(-im + inq)\frac{1}{rI}\frac{\partial}{\partial r} rR_0\left(\frac{r}{R_0} + \Delta'\right)\frac{\partial}{\partial r}\frac{1}{qR_0}(-i(m\pm 1) + inq)\phi_{m\pm 1} +
$$

$$
-\sum_\pm \frac{I}{2qR_0^2}(-im + inq)\frac{1}{rI}\frac{\partial}{\partial r} rR_0\frac{\partial}{\partial r}\frac{1}{qR_0}\frac{r}{R_0}(-i(m\pm 1) + inq)\phi_{m\pm 1} \tag{C.22}
$$

Express this term via the parallel wave vector to obtain

$$
\text{R.H.S. First Term}_m =
$$

$$
= -\frac{k_{\|m}}{rR_0}\frac{\partial}{\partial r} rR_0\frac{\partial}{\partial r} k_{\|m}\phi_m
$$

$$
+\frac{1}{rR_0}\frac{r}{R_0} k_{\|(m+1)}\frac{\partial}{\partial r} rR_0\frac{\partial}{\partial r}\left(k_{\|(m+1)}\phi_{m+1} + k_{\|(m-1)}\phi_{m-1}\right) +
$$

$$
+\frac{1}{2rR_0} k_{\|m}\frac{\partial}{\partial r} rR_0\frac{\partial}{\partial r}\left(k_{\|(m+1)}\phi_{m+1} + k_{\|(m-1)}\phi_{m-1}\right) +
$$

$$
-\frac{1}{rR_0} k_{\|m}\frac{\partial}{\partial r} rR_0\left(\frac{r}{R_0} + \Delta'\right)\frac{\partial}{\partial r}\left(k_{\|(m+1)}\phi_{m+1} + k_{\|(m-1)}\phi_{m-1}\right) +
$$

$$
+\frac{1}{2rR_0} k_{\|m}\frac{\partial}{\partial r} rR_0\frac{\partial}{\partial r}\left(k_{\|(m+1)}\phi_{m+1} + k_{\|(m-1)}\phi_{m-1}\right). \tag{C.23}
$$

Representing the other three terms, we arrive at the complete bending energy term:

$$-k_{\|m}^2 \frac{\partial^2}{\partial r^2} \phi_m + \sum_{\pm} \left(\frac{r}{R_0} k_{\|m\pm1}^2 \frac{\partial^2}{\partial r^2} \phi_{m\pm1} - \Delta' k_{\|m} k_{\|m\pm1} \frac{\partial^2}{\partial r^2} \phi_{m\pm1} \right) + \frac{m^2}{r^2} k_{\|m} \phi_m +$$

$$+ \sum_{\pm} \left(-\frac{m^2}{r^2} \frac{r}{R_0} k_{\|m\pm1}^2 \phi_{m\pm1} - \frac{m^2}{r^2} \left(\frac{r}{R_0} + \Delta' \right) k_{\|m} k_{\|m\pm1} \phi_{m\pm1} \right) \tag{C.24}$$

The inertial term is analysed in a similar manner:

$$\nabla \cdot \left(\frac{\omega^2}{v_A^2} \nabla_\perp \phi \right) = -\nabla \cdot \left(\frac{\omega^2}{v_A^2} \left(\frac{R}{R_0} \right)^2 \left(\frac{1}{B^2} \boldsymbol{B} \times \nabla \phi \right) \right) =$$

$$= \left(\nabla \cdot \boldsymbol{A} = \frac{1}{J} \frac{\partial}{\partial \xi^i} \left(J A^i \right) \right) = -\frac{1}{J} \frac{\partial}{\partial \xi^i} \left(\frac{\omega^2}{v_A^2} \left(\frac{R}{R_0} \right)^2 \frac{J}{B^2} \nabla \xi^i \cdot \left(\boldsymbol{B} \times \left(\boldsymbol{B} \times \nabla \phi \right) \right) \right) \tag{C.25}$$

The m-th Fourier component of this term is

$$\frac{1}{r} \frac{\partial}{\partial r} \left(r \Omega^2 \frac{\partial \phi_m}{\partial r} \right) - \sum_{\pm} \frac{1}{r} \frac{r}{R_0} \frac{\partial}{\partial r} \left(r \Omega^2 \frac{\partial}{\partial r} \phi_{m+1} \right) +$$

$$+ \sum_{\pm} \frac{1}{r} \frac{\partial}{\partial r} \left(r \Omega^2 \left(\Delta' + 2 \frac{r}{R_0} \right) \frac{\partial}{\partial r} \phi_{m+1} \right) - \frac{m^2 \Omega^2}{r^2} \phi_m - \sum_{\pm} \frac{m^2 \Omega^2}{r^2} (\Delta') \phi_{m+1} \tag{C.26}$$

The sum of the bending energy term and the inertial term gives for two coupled harmonics the following equations coupled through toroidicity:

$$\frac{\partial}{\partial r} \left(\frac{\omega^2}{v_A^2} - k_{\|m}^2 \right) \frac{\partial \phi_m}{\partial r} - \frac{m^2}{r^2} \left(\frac{\omega^2}{v_A^2} - k_{\|m}^2 \right) \phi_m + \frac{\varepsilon}{4 q^2 R_0^2} \frac{\partial^2 \phi_{m-1}}{\partial r^2} = 0$$

$$\frac{\partial}{\partial r} \left(\frac{\omega^2}{v_A^2} - k_{\|m-1}^2 \right) \frac{\partial \phi_{m-1}}{\partial r} - \frac{m^2}{r^2} \left(\frac{\omega^2}{v_A^2} - k_{\|m-1}^2 \right) \phi_{m-1} + \frac{\varepsilon}{4 q^2 R_0^2} \frac{\partial^2 \phi_m}{\partial r^2} = 0$$

$$\varepsilon = 2 \left(\frac{1}{4} + 1 \right) \frac{r}{R_0} = \frac{5}{2} \frac{r}{R_0} \tag{C.27}$$

These equations are the starting point equations (6.3) and (6.4) from Chapter 6.

C.2   RADIATIVE DAMPING OF TAE AND KINETIC TAE

For simplicity, we consider modes with high poloidal mode number, $m \gg 1$, in a torus with circular magnetic flux surfaces, low magnetic shear, and small inverse aspect ratio, $\frac{r}{R} \ll 1$. In this limiting case, each eigenmode is highly localised in the vicinity of the centre of TAE-gap and consists of two neighbouring poloidal harmonics $\varphi_m(r)$, $\varphi_{m-1}(r)$. With the first-order finite Larmor radius included in the vicinity of the TAE-gap, the appropriate set of equations for the m-th and $(m-1)$-th amplitudes of the electrostatic potential is given by the following coupled fourth-order differential equations [6.11,6.17]:

$$\rho^2 \frac{\mathrm{d}^4}{\mathrm{d}r^4}\varphi_m + L_m\varphi_m + \varepsilon\frac{1}{4q^2R^2}\frac{\mathrm{d}^2}{\mathrm{d}r^2}\varphi_{m-1} = 0,$$

(C.28)

$$\rho^2 \frac{\mathrm{d}^4}{\mathrm{d}r^4}\varphi_{m-1} + L_{m-1}\varphi_{m-1} + \varepsilon\frac{1}{4q^2R^2}\frac{\mathrm{d}^2}{\mathrm{d}r^2}\varphi_m = 0,$$

where $\varepsilon = \left(\frac{5}{2}\right)\left(\frac{r}{R}\right)$, the differential operator L_m is defined as

$$L_m\varphi_m \equiv \frac{\mathrm{d}}{\mathrm{d}r}\left(\frac{\omega^2}{V_A^2} - k_{\|m}^2\right)\frac{\mathrm{d}\varphi_m}{\mathrm{d}r} - \frac{m^2}{r^2}\left(\frac{\omega^2}{V_A^2} - k_{\|m}^2\right)\varphi_m,$$

(C.29)

and the first-order Larmor radius term is defined as

$$\rho^2 \equiv \frac{\rho_i^2}{4R_0^2q_0^2}\left(\frac{3}{4} + \frac{T_e}{T_i}\left(1 - i\delta_e\right)\right).$$

(C.30)

For solving (C.28), we follow the TAE-ansatz and consider "outer" region away from the TAE-gap and "inner" layer surrounding the TAE-gap. The outer region has only cylindrical part of equations similar to (6.18) and (6.19). In the outer region, we neglect both toroidicity and non-ideal effects and solve it similarly to TAE. In contrast to the outer region, the vicinity of the TAE-gap has both small parameters, the toroidal coupling, and the non-ideal Larmor radius terms, playing an important role as the solutions of (C.28) in the inner layer are highly peaked and the small parameters mentioned are in front of the terms with the high-order derivatives. We keep the terms of (C.28) with high-order derivatives to obtain the following equations for the inner layer [6.11,6.17]:

$$\lambda^2 \frac{\mathrm{d}^4}{\mathrm{d}z^4}\varphi_m + \frac{\mathrm{d}}{\mathrm{d}z}\left[(g+z)\frac{\mathrm{d}\varphi_m}{\mathrm{d}z}\right] + \frac{\mathrm{d}^2}{\mathrm{d}z^2}\varphi_{m-1} = 0,$$

(C.31)

$$\lambda^2 \frac{\mathrm{d}^4}{\mathrm{d}z^4}\varphi_{m-1} + \frac{\mathrm{d}}{\mathrm{d}z}\left[(g-z)\frac{\mathrm{d}\varphi_{m-1}}{\mathrm{d}z}\right] + \frac{\mathrm{d}^2}{\mathrm{d}z^2}\varphi_m = 0,$$

where the normalized frequency g is given by (6.13), the dimension-less radial variable z is given by (6.16), and the non-ideal parameter λ^2 is given by (6.31). The coupled equations (C.31) can be integrated once to give

$$\lambda^2 \frac{\mathrm{d}^2}{\mathrm{d}z^2}U + (g+z)U + V = C_m,$$

(C.32)

$$\lambda^2 \frac{\mathrm{d}^2}{\mathrm{d}z^2}V + (g-z)V + U = -C_{m-1}.$$

Here, we introduced $U \equiv \frac{\mathrm{d}\varphi_m}{\mathrm{d}z}, V \equiv \frac{\mathrm{d}\varphi_{m-1}}{\mathrm{d}z}$, and the integration constants C_m, C_{m-1} have to be chosen so that at $z \to \infty$ the asymptotes of U, V match onto the "outer" solutions of (6.24) at $\frac{|x|}{S} \ll 1$. The order of the inner layer equations (C.32) can be further reduced by Fourier transformation (C.32) and solving the problem in the k_r-space [6.15–6.17]. Denoting the respective Fourier transforms of $U(z), V(z)$ with $u(k), v(k)$, we obtain

$$L_k u + v = 2\pi C_m\delta(k),$$

(C.33)

$$L_{-k}v + u = -2\pi C_{m-1}\delta(k),$$

where

$$L_k \equiv i\frac{\mathrm{d}}{\mathrm{d}k} + g - \lambda^2 k^2. \tag{C.34}$$

Eqs. (C.33) and (C.34) are to be solved with the boundary conditions $u(k) \to \infty$, $v(k) \to \infty$ as $k \to \infty$, and the jump conditions at the origin, which result from the δ-functions in (C.33):

$$u(0^+) - u(0^-) = -2\pi i C_m,$$

$$v(0^+) - v(0^-) = -2\pi i C_{m-1}. \tag{C.35}$$

Away from $k=0$ (that corresponds to the MHD outer region), (C.33) and (C.34) are combined to give two uncoupled second-order differential equations for $u(k)$, $v(k)$:

$$[L_{-k}L_k - 1]u = 0,$$

$$[L_k L_{-k} - 1]v = 0, \tag{C.36}$$

which in the limit of moderate k, $k \le \lambda^{-1}$ takes the form of a Schrodinger-type equation

$$\frac{\mathrm{d}^2}{\mathrm{d}k^2}u + (E - V(k))u = 0, \tag{C.37}$$

where $E \equiv -(1 - g^2)$ is the eigenvalue associated with the mode eigenfrequency, and

$$V(k) \equiv 2g\lambda^2 k^2 - \lambda^4 k^4 \tag{C.38}$$

is the effective potential for the Schrodinger problem in the k-space.

The most important part of (C.38) is the first (quadratic) term, the sign of which depends on the sign of the normalized frequency g. In accordance with Figure 6.2 (right), we have $g < 0$ for the mode frequencies below the centre of TAE-gap frequency ω_0, and $g > 0$ – for the mode frequencies above ω_0. Consequently, the quadratic term of (C.38) gives a hill of the potential for the modes below ω_0, and a potential well – for the modes above ω_0. Figure C.1 illustrates the form of the potential (C.38) for two reference values, $g = \pm 1$. This is how the asymmetry in KAWs between the

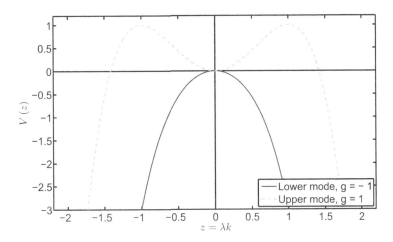

FIGURE C.1   Effective potential (C.38) for modes at the bottom tip, $g = -1$, and at the top tip, $g = +1$, of the TAE-gap.

frequency regions above and below the central frequency of the TAE-gap $\omega_0 = 1/(2qR)$ manifests itself in the k-space.

The problem of the radiative damping, in terms of the Schrodinger equation approach in the k-space we developed here, is to investigate what fraction of u tunnels through the potential $V(k)$ due to the non-ideal Larmor radius effects. The usual TAE mode obtained in ideal MHD approach with its eigenvalue (6.14) and (6.25) corresponds to the potential well of the $\delta(k)$-function type in (C.33), (C.34) taken in the absence of the non-ideal effects, $\lambda=0$. Radiative damping for TAE corresponds to the tunnelling through the potential barrier from $k=0$ to the turning points,

$$k_\pm = \pm\frac{\sqrt{1+g}}{\lambda}. \tag{C.39}$$

After tunnelling, the radiative wave matches the outgoing wave solution

$$u(k) \sim \exp\left[-i\frac{\lambda^2 k^3}{3}\right], \, k \to \infty \tag{C.40}$$

which is the large-k WKB solution of (C.36). For TAE at the bottom of the TAE-gap, the asymptotic matching procedure gives the value of the radiative damping (6.30).

Near the upper boundary of the gap, $g \approx +1$, the non-ideal effects provide a local well. The asymptotic analysis in this case shows an existence of a quasi-stationary discrete spectrum (6.32) of kinetic TAE modes inside this potential well. Frequencies of KTAEs are just above the gap, and these modes have a relatively low radiative damping (6.33).

Appendix D
Diagnostics of Alfvén Eigenmodes Excited by Energetic Particles

Because energetic particles excite instabilities via wave-particle resonances, excited shear Alfvén eigenmodes have their frequencies close to the poloidal and toroidal characteristic frequencies of drift orbits of the energetic particles driving the modes. For present-day tokamaks, AEs of various types, from the fishbones to EAEs and NAEs, cover the frequency range from ~10 to ~500 kHz in the plasma reference frame. The best tested and most common instrument for detecting electromagnetic modes in this frequency range is a magnetic sensor, most often a Mirnov coil, measuring time-dependent magnetic flux just outside the plasma. A set of toroidally and poloidally separated Mirnov coils can also provide information on the mode numbers and the directivity of propagation of unstable AEs. For an in-depth study of any particular type of AE, and especially for using AEs detected as MHD markers for providing information on plasma equilibrium properties (Alfvén spectroscopy discussed in Chapter 10), a knowledge of some specific properties of AEs is required. In this Appendix, we consider in detail the properties of TAE and describe, as an example, the set of Mirnov coils on JET.

However, the use of magnetic sensors in future burning DT plasma experiments, such as ITER, could be difficult [D.1]. To be protected in the harsh environment with high flux of DT neutrons, the magnetic sensors must be hidden well, for example, behind a thick blanket. Moreover, the plasma itself will be much larger, and detection of AEs deep in the plasma core cannot be guaranteed. Therefore, alternative techniques of AE detection, which are more compatible with DT operation, are required.

Microwave diagnostics, reflectometry, and interferometry measuring fluctuations of plasma density associated with AE are described in this Appendix as the possible avenues for development in view of a large-scale DT plasma. In particular, X-mode reflectometry was used successfully for detecting alpha particle-driven AEs in TFTR [D.2], which has recently shown its potential in obtaining information on $q(R, t)$ from the positions of AEs [D.3], while the FIR interferometry with vertical line-of-sight through the magnetic axis has become the most reliable diagnostic technique for detecting AEs on the magnetic axis of JET [D.4].

Diagnostics measuring fluctuations of electron temperature associated with AE, ECE, and soft X-ray (SXR) are other options to consider. Excellent performance of ECE diagnostics on present-day machines, especially on DIII-D [D.5] and ASDEX-Upgrade [D.6], has led to top-quality internal measurements of AEs, providing an opportunity for a complete theory-to-experiment successful comparison [D.7].

We note here that (1) magnetic sensors do not provide information on the *amplitude* of an AE perturbation inside the plasma, while X-mode reflectometry and ECE do, and (2) toroidally and/or poloidally separated magnetic sensors provide information on the *mode numbers*, while interferometry, reflectometry, ECE, and SXR do not, and even the *directivity of the wave propagation*, which comes naturally from the magnetics requires special arrangements for reflectometry. The diagnostics above are complementary to each other, and when they are used on present-day machines in a combination, see, for example, [D.8], they can provide all the information on AEs needed for further interpretation and modelling. The problem now is to extend such complementarity options to DT burning plasma experiments.

D.1 PROPERTIES OF TAE

The frequency gap in the Alfvén continuum appears in toroidal geometry at frequency

$$\omega = k_{\|m}(r)V_A(r) = -k_{\|m+1}(r)V_A(r), \tag{D.1}$$

so that at the radius associated with TAE-gap we have

$$k_{\|m}(r) = -k_{\|m+1}(r), \tag{D.2}$$

that is

$$\frac{1}{R_0}\left(n - \frac{m}{q(r)}\right) = -\frac{1}{R_0}\left(n - \frac{m+1}{q(r)}\right). \tag{D.3}$$

Expression (D.3) shows that at the TAE-gap the safety factor is related to TAE toroidal and poloidal mode numbers via

$$q(r) = \frac{m+1/2}{n}. \tag{D.4}$$

By substituting this value of TAE-specific safety factor in (D.1), we obtain the characteristic frequency of the TAE-gap,

$$\omega_0 = \frac{V_A(r_0)}{2R_0 q(r_0)}, \tag{D.5}$$

where r_0 is the radius of the TAE-gap. It is important to note here that TAE-gap frequency (D.5) does not depend on either toroidal or poloidal mode numbers of TAE.

The relation (D.4) implies that the very existence of a TAE-gap (and relevant weakly damped TAE-mode inside it) in toroidal plasma depends on whether a magnetic flux surface with the relevant value of q exists in the plasma. Table D.1 shows the values of q required for TAEs with poloidal and toroidal mode numbers within 1, ..., 6.

We now estimate from (D.5) the characteristic frequencies of TAE-gap for typical JET parameters $B_0 \cong 3T$; $n_i = 5 \times 10^{19}$ m^{-3}; $m_i = m_D$, giving $V_A \cong 6.6 \times 10^6$ m/s. For typical value of $q = 1.1$ we obtain the frequency estimate:

$$\omega_0 \cong 10^6 \text{ s}^{-1} \rightarrow f_0 \equiv \omega_0 / 2\pi \cong 160 \text{ kHz}. \tag{D.6}$$

This frequency is estimated for the plasma reference frame. However, plasma often rotates, and a proper correction for the relevant Doppler shift of the mode frequency is needed. For example, uni-directional NBI on JET shown in Figure D.1 spins up the plasma toroidally, and the toroidal plasms

TABLE D.1
q-values of TAE-Gaps Determined by (D.4) for $0.5 \leq q \leq 6.5$

	$m=1$	$m=2$	$m=3$	$m=4$	$m=5$	$m=6$
$n=1$	1.5	2.5	3.5	4.5	5.5	6.5
$n=2$	0.75	1.25	1.75	2.25	2.75	3.25
$n=3$	0.5	0.83	1.167	1.5	1.83	2.167
$n=4$		0.625	0.875	1.125	1.375	1.625
$n=5$		0.5	0.7	0.9	1.1	1.3
$n=6$		0.583	0.75	0.917	1.083	

rotation may be about $f_{\text{rot}}(r) \sim 10 \div 25$ kHz (maximum achieved was ~40 kHz) depending on the power of NBI. Frequencies of AEs with mode number n in laboratory reference frame, f_n^{LAB}, and in the plasma, f_n^0, are related through the Doppler shift $n f_{\text{rot}}(r)$:

$$f_n^{\text{LAB}} = f_n^0 + n f_{\text{rot}}(r). \tag{D.7}$$

In the case of several TAEs with neighbouring toroidal mode numbers excited at once in toroidally rotating plasmas, the frequency separation between them is approximately $f_{\text{rot}}(r)$:

$$f_{n+1}^{\text{LAB}} - f_n^{\text{LAB}} = \left(f_{n+1}^0 - f_n^0\right) + \left((n+1) - n\right) f_{\text{rot}}(r) \approx f_{\text{rot}}(r). \tag{D.8}$$

where we used the independence of TAE-gap frequency (D.5) on n,

$$f_{n+1}^0 \approx f_n^0. \tag{D.9}$$

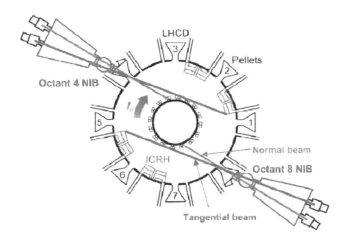

FIGURE D.1   Geometry of NBI injection system on JET (view from the top of the machine).

D.2   DIAGNOSING AE WITH MIRNOV COILS

It was shown in the description of TAE, Eqs. (6.36) and (6.37) in Chapter 6, that only magnetic perturbations perpendicular to the equilibrium magnetic field, δB_r and δB_ϑ, are significant in TAE, and $\delta B_r \ll \delta B_\vartheta$ if the mode structure satisfies $\partial/\partial r \gg m/r$. Similar relations are valid for a majority of shear Alfvén eigenmodes (EAEs, NAEs, ACs). For detecting AE perturbations, magnetic coils mounted just outside plasma are employed on a majority of magnetic fusion devices. Figure D.2 shows the geometry of Mirnov coils on JET, which measure an oscillatory magnetic flux induced by δB_ϑ passing through the coil axis. The coils provide the value of

$$\frac{\partial}{\partial t} \delta B_\vartheta^{\text{edge}} \cong \omega \cdot \delta B_\vartheta^{\text{edge}} \tag{D.10}$$

Owing to the high values of TAE frequency, $\omega \cong 10^6 \, \text{s}^{-1}$, which is a factor in front of the perturbed field $\delta B_\vartheta^{\text{edge}}$ in (D.10), the coils are sensitive enough to detect the perturbed magnetic fields as low as $\left|\delta B_\vartheta^{\text{edge}} / B_0\right| \cong 10^{-8}$. The data acquisition system of the coils typically uses the sampling rate of 1 MHz on JET, so measurements of AEs up to 500 kHz can be made. The Mirnov coils are calibrated, that is, they provide the same amplitude and phase response to the same test signal, and hence, the cross-analysis of the coils separated in toroidal direction provides an accurate information regarding the toroidal mode numbers.

To determine the toroidal mode number n of a mode, two or more toroidally separated Mirnov coils are used and the phase shift is measured between them as Figure D.3 illustrates. Figure D.3 shows a toroidal plasma with the directivities indicative of the magnetic fields, plasma current, the beam injection, and the toroidal projections of electron and ion diamagnetic frequencies. Let us assume we have two toroidally separated magnetic pick-up coils at toroidal angles φ_1 and φ_2. If both coils measure at the same time t_0, the same sinusoidal wave of frequency ω_0, then a phase shift α exists between the measurements due to the finite toroidal angle, $\Delta\varphi$, distance between the

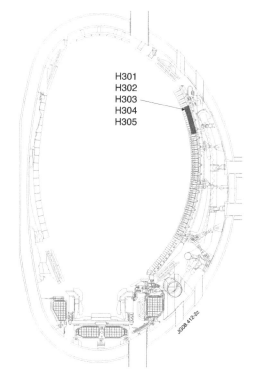

FIGURE D.2 JET cross-section showing the position and poloidal directivity of five high-frequency Mirnov coils H301 … H305 separated in the toroidal angle.

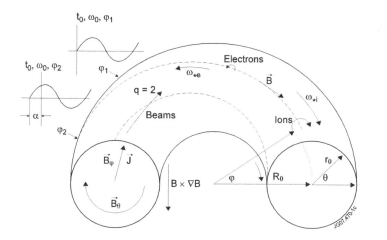

FIGURE D.3 Sinusoidal signals measured at different toroidal angles at the same time and at same frequency are shifted in phase by α.

coils. For a known $\Delta\varphi$ and the measured α, one can calculate what part of the wavelength fits within the toroidal distance between the probes, and an extrapolation of this length to a complete toroidal circle gives the toroidal mode number n.

D.3   FURTHER ADVANCES IN DIAGNOSING ALFVÉN INSTABILITIES

Further advances in diagnosing Alfvén instabilities are associated with recent expansion of tools and techniques for the detection and identification of unstable modes. Reflectometry, interferometry, ECE, and SXR measurements of perturbed electron density and temperature associated with AEs were successful alternatives to magnetic sensors.

In toroidal geometry, the perturbed electron density caused by AEs is found from the continuity equation:

$$\frac{\delta n}{n_0} = -\xi \cdot \frac{\nabla n_0}{n_0} - \nabla \cdot \xi = \left(\frac{\hat{n}}{L_n} - 2\frac{\hat{R}}{R^2} \right) \cdot \xi, \tag{D.11}$$

where δn, n_0 are the perturbed and equilibrium densities, ξ is the plasma displacement caused by the perturbed electric field via $-i\omega\xi = \left(c / B_0^2\right)\left[\delta E \times B_0\right]$, and L_n is the radial scale length of the density profile. The first term in the right-hand side of (D.11) describes the usual convection of plasma proportional to the gradient of equilibrium density. The second term $\propto 1/R$ in (D.11) is caused by the toroidicity and causes a non-zero δn even if the profile of n_0 is flat. This term also causes an anti-ballooning structure of the density perturbations δn even when δB has no significant in-out asymmetry [D.9].

A launched microwave O-mode beam on JET with frequency above the cut-off frequency of O-mode was found to deliver detection of AEs far superior to that made with magnetic sensors [D.10]. This "O-mode interferometry" shows many more unstable AEs in the plasma core, some

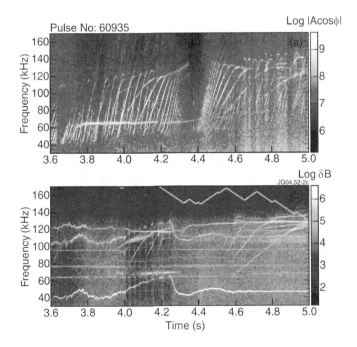

FIGURE D.4   Spectrograms showing AEs with different toroidal and poloidal mode numbers: Top: interferometry O-mode measurements with 45.2 GHz microwave beam. Bottom: measurements of same modes with Mirnov coils.

of which are not even detected with Mirnov coils as Figure D.4 illustrates. The particular settings of the O-mode system on JET does not allow measurements to be made above plasma densities of ~6–7 10^{19} m^{-3}, so a higher frequency instrument had to be developed.

The standard far infra-red (FIR) JET interferometer was digitised to a high sampling rate, which enabled detecting AEs in plasmas of high density [D.4]. The FIR interferometer on JET has four vertical lines-of-sight as Figure D.5 shows, which provide measurements of the line-integrated perturbed density perturbations. The FIR interferometry often detects AEs deep in the plasma core, which are hardly seen with Mirnov coils, and the quality of the interferometry signal is quite

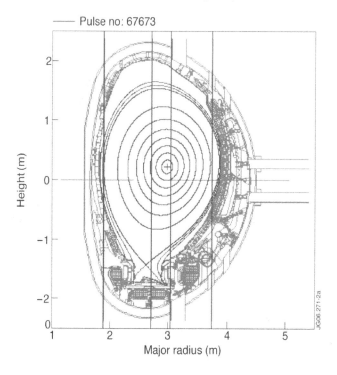

FIGURE D.5 Geometry of JET interferometer with vertical lines-of-sights (counted from left to right as Channels 1 … 4)

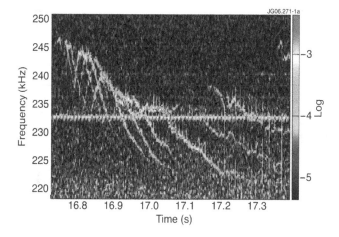

FIGURE D.6 Core-localised TAEs inside the $q = 1$ radius (tornado modes) detected with the vertical Channel 3 of the JET interferometer shown in Figure D.5 (line-of-sight passing through the magnetic axis).

satisfactory as Figure D.6 shows. A similar FIR interferometry technique was employed for detecting AEs in DIII-D discharges. It was observed for the first time that a "sea of modes" exists in reversed-shear DIII-D, with toroidal mode numbers up to $n = 40$ [D.11]. The interferometry technique has significantly increased the quality of AE detection and assures that all unstable AEs are detected even deeply in the plasma core. As the interferometry technique of detecting AEs requires only interferometers used for plasma density measurements, this method is a good candidate for ITER and DEMO.

The main limitation of using interferometry or Mirnov coils for detecting AEs is that the AEs cannot be localised from the measurements and the amplitudes of AEs cannot be found with precision. However, the successful development of ECE [D.5] and ECE imaging [D.6], beam emission spectroscopy (BES) [D.12], and phase contrast imaging (PCI) [D.13] have addressed the problem of measuring mode structure. Together with the existing SXR technique and X-mode reflectometry used for observing alpha-driven AEs in DT plasmas [D.2], the new diagnostics provide measurements of the spatial structure of the modes to a degree required for an accurate experiment-to-theory comparison.

REFERENCES

[D.1]   Progress in the ITER Physics Basis, *Nucl. Fusion* **47** (2007) S1.
[D.2]   R. Nazikian et al., *Phys. Rev. Lett.* **78** (1997) 2976.
[D.3]   S. Hacquin et al., *Plasma Phys. Control. Fusion* **49** (2007) 1371.
[D.4]   S.E. Sharapov et al., *Nucl. Fusion* **46** (2006) S868.
[D.5]   M.A. Van Zeeland et al., *Phys. Rev. Lett.* **97** (2006) 135001.
[D.6]   S.J. Freethy et al., *Rev. Sci. Instrum.* **87** (2016) 11E102.
[D.7]   W.W. Heidbrink et al., *Phys. of Plasmas* **24** (2017) 056109.
[D.8]   M.A. Van Zeeland et al., *Nucl. Fusion* **46** (2006) S880.
[D.9]   R. Nazikian et al., *Phys. Rev. Lett.* **91** (2003) 125003.
[D.10]   S.E. Sharapov et al., *Phys. Rev. Lett.* **93** (2004) 165001.
[D.11]   R. Nazikian et al., *Phys. Rev. Lett.* **96** (2006) 105006.
[D.12]   R.D. Durst et al., *Phys. Fluids* **B4** (1992) 3707.
[D.13]   E.M. Edlund et al., *Plasma Phys. Control. Fusion* **52** (2010) 115003.

Index